AUTISM, MR TRUMP

Othmane Boumaalif, MD

AUTISM, MR TRUMP

Beyond Tylenol and Leucovorin - A Guide to a New Understanding

ISBN numérique : 9791040593034
EAN papier : 9791040593041

Librinova

Courriel : contact@librinova.com
Internet : www.librinova.com

Contents

Disclaimer ... 9
Preface ... 11
Autism: Toward a New Paradigm ... 13
A Runaway Epidemiology: Autism in Numbers ... 21
Beyond Diagnostic Labels: Rethinking Autism ... 25
Reading Leo Kanner's First Article on Autism ... 29
Stop "Psychiatrizing" Autism: It Is a Systemic Disease ... 33
Autism's Inflammatory Nature ... 37
Autism and Autoimmunity: A Hidden Connection ... 39
Autism and Mast Cell Activation ... 43
Autistic Regression: A Controversial Concept ... 47
Autism as a Mitochondrial Disease ... 51
The Genetic Puzzle of Autism ... 57
Autism and Genetic Variations: Understanding SNPs ... 61
Genetic Polymorphisms and Autism: The Case of MTHFR ... 65
Autism and the So-Called "Junk DNA" ... 69
Autism and the Forgotten Organ: The Gut Microbiome ... 73
p-Cresol and Propionic Acid: The Gut-Brain Connection in Autism ... 79
The Fungal Hypothesis in Autism ... 81

The Brain-Gut Connection and the Vagus Nerve in Autism 85
Autism and the Thyroid Connection 89
Carnitine and Autism: A Hidden Connection 91
Paracetamol and Autism: A Controversial Link 95
Autism and Calcium: A Hidden Connection 101
Autism and Microglia 105
Vitamin D: The Missing Piece of the Autism Puzzle 109
The Role of GABA and Glutamate Systems in Autism 115
Serotonin, Dopamine, and Autism 119
Autism and the Kynurenine Pathway 125
PANS and PANDAS: When the Immune System Shapes the Mind 127
Autism and Aggression: Understanding the Hidden Triggers 131
Autism and Language: Cracking the Communication Code 135
Beyond Every Protocol: Understanding Autism's Complexity 139
Mega-Analysis and Clinical Research on Autism 143
Translational Medicine: A Therapeutic Hope for Autism145
Preventing Autism Is Possible 149
Fecal Transplants and Autism: What the New Science—and New Rules—Really Say 153
The Future Perspective: The Cell Danger Response Theory 157
Future Perspectives: Transcranial Magnetic Stimulation.. 161
Stem Cell Therapy: Between Hope and Illusion 165

Advocacy for Early Autism Diagnosis 167
For a Truly Inclusive School 171
Beyond Autism: Uncovering Hidden Metabolic and Genetic Disorders 175
Towards a Comprehensive Theory: The Seven Breaches Model 179
Epilogue 183
Further Reading 185

Disclaimer

This book is intended to share information based on scientific research and personal experience.
However, it is not a substitute for professional medical advice, diagnosis, or treatment.
For any questions regarding your health or the health of your loved ones, you should always seek the guidance of a physician or qualified healthcare professional. The author and publisher cannot be held responsible for any consequences resulting from the improper use of the information contained in this book.
This book is an independent work of analysis and commentary. It is not authorized, sponsored, or endorsed by Donald J. Trump, his family, or any related organization. Any reference to public figures is made solely for informational, educational, and critical purposes under the principles of freedom of expression and fair use. The opinions expressed herein are exclusively those of the author.

Preface

On September 22, 2025, U.S. President Donald Trump stood at the White House, alongside Health and Human Services Secretary Robert F. Kennedy Jr. and senior health officials, to address the nation about autism. It was no ordinary press conference. For the first time at a presidential level, autism was recognized as an epidemic of immense scale, requiring urgent national and global attention.

The announcements were bold. President Trump declared that autism is not only a medical challenge but also a societal one. He announced new measures: an updated label for leucovorin (folinic acid) to support children with cerebral folate deficiency linked to autism, and new guidance warning against widespread use of acetaminophen (Tylenol) during pregnancy. For families around the world, it felt like a turning point.

This moment carried undeniable positives. Autism was finally being spoken of outside the narrow frameworks of pure genetics or psychoanalysis. For the first time, biomedical and immune factors were placed at the center of the conversation. By highlighting leucovorin, the administration indirectly acknowledged the possible autoimmune dimension of autism. For countless families, that recognition alone was long overdue.

Yet concerns remain. True to his populist style, President Trump reduced autism's complex causality to a single factor: acetaminophen. While there is research suggesting a link, such a simplistic explanation risks obscuring the broader reality. Autism is not the result of one medication. It is the outcome of multiple overlapping vulnerabilities: immune dysregulation, metabolic imbalance, gut-brain disruption, mitochondrial stress, environmental toxins, and chronic inflammation. To reduce it to Tylenol alone is misleading.

Leucovorin, too, while promising in carefully selected cases, is not a miracle cure. It has helped some children with cerebral folate deficiency improve in speech and social skills, but it is no universal solution. Its use must be cautious, personalized, and combined with therapies addressing the whole child.

I write these words not from Washington, but from Morocco. My journey began when my own son was diagnosed with autism at just two years old. That moment transformed my life and medical career. Since then, I have worked with thousands of children and families across Morocco, North Africa, Europe, the Middle East, and beyond. I have seen their struggles—the sleepless nights, the gut problems, the self-injury, the isolation—and I have witnessed their resilience. Autism is not an abstract concept to me. It is a lived reality.

This book is my contribution to a new, science-driven, compassionate movement. We must go beyond politics, beyond borders, and beyond ideology. Autism is a global challenge. It demands unity, knowledge, and courage. No nation can solve it alone. Families from all around the world, are waiting for real solutions, not soundbites.

For Iyad, my son. For every child and every parent walking this difficult path. May this book open the door to hope, healing, and a better future.

Autism, Mr Trump is not a provocation – it is a call to awareness.
Through this title, science speaks to power and to every parent, urging a new, systemic understanding of autism – one rooted in biology, compassion, and hope.

Autism: Toward a New Paradigm

"You have to grieve this child." Those words, spoken by a child psychiatrist, struck my wife and me like a blade. That day, we learned that our son—only two years old—had just been diagnosed with autism. The sentence, brutal and implacable, rang out like a final verdict, asking us to give up any hope of seeing him progress or flourish like other children. The shock was immense, leaving a deep wound within us.

Nothing seemed to predestine our family for such an ordeal. We lived in a loving, stable home. My wife had put her career on hold for two years, wanting to offer our only child the best conditions in which to grow. But destiny—unpredictable and cruel—decided otherwise.

This diagnosis did not only overturn our daily life; it set off an unrelenting quest to understand and to act. What at first looked like a desperate attempt to save our child's future became a personal and professional journey that redefined my life.

I am a general practitioner, yet my career has never been limited to a single discipline. Over the years, I have explored many areas of medicine. I specialized in systemic diseases, geriatrics, diabetology, nutrition, as well as occupational medicine and toxicology. My long path through the health industry allowed me to train in fields such as biostatistics, pharmaco-epidemiology, clinical research, and even more technical domains, ranging from molecular screening to translational research.

With hindsight, I see these stages as a "divine" preparation—a path traced to arm me with the tools needed to grasp the complexity of autism and to help my son. I am profoundly a person of faith, and that faith allowed me to accept this challenge as a mission.

This journey led me far beyond my family circle. What was once a modest medical practice became a place of hope for thousands of families. Autistic children came from all over Morocco and even from abroad—from France, Spain, Italy, Luxembourg, the United Arab Emirates, Saudi Arabia, Canada, and the United States—and found a welcome there.

Every child, every parent, carried a similar look: a blend of pain and hope, a universal wound I knew in my own flesh. What I learned while accompanying them nourished my fight for my son and also strengthened my commitment to this community—so often marginalized and misunderstood.

Over time, a conviction took hold in me: autism is not simply a "difference" or an "alternative way of functioning," as some romanticized discourses try to present it. Behind autism there is immense suffering. Autistic people are not always individuals endowed with extraordinary talents or charming quirks. The reality is far more cruel.

I think of children who are totally nonverbal, unable to communicate. Of those who do not sleep; of debilitating intestinal disorders; of motor difficulties; of the absence of compassion; and of repetitive, compulsive behaviors. I also think of self-injury, anxiety attacks, aggression, violent temper outbursts, and emotional collapses that punctuate their daily lives—and those of their families.

These families, often left to themselves, find themselves isolated, neglected by their loved ones, and sometimes judged or stigmatized by society. That society, so often cruel, passes hasty judgments, sharp criticisms, and rarely extends a helping hand. Added to this are the inertia and misunderstanding of the medical profession, and the lack of any truly inclusive vision in the school system.

Faced with this picture, my approach has always been simple—but unique. I do not claim to wield a magic wand or to possess a miracle cure. My approach rests on translational medicine: I use tools grounded in the most recent research, combining basic science and clinical data to offer a response that is tailored and personalized to each child.

Thanks to this approach, I have had successes. I am proud to have contributed to hundreds of complete recoveries from autism. Other children have shown significant progress. But I have also known failures. There is no miracle. Every child is unique, and outcomes depend on many factors, often beyond our control.

What makes my work special is the bond I establish with families. This positive transfer—based on love and trust—has allowed me to give the best of myself to every child. Through them, I always saw my son, which amplified my determination.

This suffering allowed me to find a mission: to accompany thousands of other children and their families. The quest continues. It is demanding, sometimes exhausting, but also deeply rewarding.

This colossal and exhausting work would never have been possible without the love and tenacity of my wife, Meryem. In the darkest moments, she drew upon an inexhaustible energy to carry on this fight—our fight. Through her, I pay tribute to all the mothers of autistic children who endure everything with unshakeable strength, sustained by the hope of seeing their children flourish.

This struggle is not romantic. It is brutal and cruel. It includes scenes that many families live through every day: a child who wakes at midnight and self-injures; another who screams or strikes his parents during endless car rides; a child who refuses to eat to the point that IV infusions must be considered. These are children who repeat stereotyped gestures for hours on end, who undress in public, or who display deeply disconcerting social behaviors. They are not "poorly raised" children or the product of parental shortcomings; they are children in pain, showing the signs of an underlying disease that goes far beyond the simple notion of neurodiversity.

This suffering touches not only the children, but their families as well. Couples break apart; parents sink into depression or live in total isolation, unable to find support or empathy.

This book is a response to that reality. It rests upon scientific evidence to demystify autism and offers a sincere call to all stakeholders—physicians, educators, families, and decision-makers—to bring our energies together and overcome this disease.

For Iyad, my son. For all the autistic children I have had the honor to accompany. For all those courageous families. I dedicate this book to you, in the hope that it opens a door toward a better life.

My approach is meant to be modest and respectful. It does not in any way call into question the competence, integrity, or dedication of my colleagues, whatever their discipline. On the contrary, it is part of a desire for complementarity, aiming simply to contribute to a better understanding of autism and to possible approaches for improving care. This work reflects above all a sincere wish to share knowledge and experiences, in the hope of offering helpful insight to the families and professionals concerned.

A systemic disease is a condition that goes beyond the simple involvement of a single organ or isolated function, and instead engages a multitude of interconnected biological mechanisms throughout the body. Autism—long perceived as an exclusively neurodevelopmental disorder—reveals itself to be a truly systemic disease, affecting diverse biological systems and operating according to a complex dynamic of interactions between the brain, immunity, metabolism, and the environment. Numerous studies highlight these multiple dysfunctions, which cannot be understood or treated from a reductionist perspective.

One of the pillars of this systemic view rests on chronic inflammatory processes, both cerebral and peripheral. Elevated levels of inflammatory cytokines—such as IL-1β, IL-17, or TNF—as well as persistent activation of microglia in the brain, alter the development of synapses and neuronal connectivity, thereby disrupting crucial functions like cognition and social interaction. This brain inflammation is often accompanied by a systemic immune imbalance. For example, an excess of Th17 lymphocytes—associated with acute inflammatory states—and a deficiency of regulatory T cells (Treg)—charged with modulating these responses—create a state of constant immune activation that affects the entire organism.

These immune disturbances also express themselves in the gastrointestinal tract, where up to 70% of autistic individuals suffer from symptoms such as intestinal dysbiosis or hyperpermeability. These conditions produce an imbalance in the bacterial flora, favoring the proliferation of pathogens like Clostridia, capable of producing neuroactive toxins such as p-cresol. These toxins modify neurotransmitter levels, notably raising dopamine to toxic levels and worsening behavioral and emotional symptoms. Moreover, alteration of the intestinal barrier allows inflammatory molecules to pass into the bloodstream, contributing to systemic inflammation that aggravates brain dysfunction through the gut-brain axis.

In parallel, metabolic and mitochondrial abnormalities are central elements in this systemic approach to autism. Mitochondria—the true energy powerhouses of cells—are often dysfunctional in autistic individuals, which results in insufficient energy production and generalized oxidative stress. These abnormalities directly affect neuronal development by making neurons more fragile and reducing their capacity to meet the growing energy demands of the developing brain. In addition, toxic compounds derived from intestinal dysbiosis—such as p-cresol and propionic acid—disrupt essential metabolic cofactors like coenzyme A, thereby worsening energy deficits and reinforcing systemic disturbances.

The environment also plays a determining role in this systemic dynamic. Prenatal exposures to toxins such as pesticides, heavy metals, or endocrine disruptors interact with genetic predispositions to alter metabolic and epigenetic pathways. For example, disturbances in DNA methylation—a key process for regulating gene expression—compromise synaptic development and brain plasticity. These environmental factors, often insidious, create lasting vulnerabilities that accumulate with immune, metabolic, and neurological imbalances, amplifying neurodevelopmental alterations.

Finally, neuroinflammation is a cornerstone of this systemic picture of autism. Microglia—immune cells residing in the brain—become excessively activated in response to various danger signals. This chronic activation leads to overproduction of inflammatory cytokines that disrupt the balance between neuronal excitation and inhibition. Although these alterations were long considered irreversible, recent discoveries about brain plasticity show that it is possible to mitigate these dysfunctions through targeted interventions, opening promising therapeutic perspectives.

Thus autism cannot be reduced to a simple brain disorder. It represents a systemic pathology in which interactions among the brain, immunity, metabolism, the gut, and the environment play out in a complex dynamic. By understanding these interconnected mechanisms, we can move beyond traditional approaches centered solely on behavioral symptoms and adopt more global and personalized strategies. Such interventions may include treatments to modulate inflammation, restore the microbiota, correct metabolic imbalances, or limit environmental exposures.

Throughout this book, I detail the systemic nature of autism and explore how these multiple dysfunctions interlock to produce the neurodevelopmental alterations we observe. My objective is to propose an integrated vision of autism—not only to better understand this condition, but also to identify avenues of intervention that could significantly improve the quality of life of autistic children and their families. By recognizing autism as a systemic disease, we open the way to innovative care strategies capable of targeting root causes rather than limiting themselves to visible manifestations.

In this book, I also present a new theoretical model for understanding the genesis of autism, based on the idea that the disorder emerges from the alignment of several biological vulnerabilities, or "breaches." Inspired by the Swiss-cheese model used in risk management, this framework proposes that autism does not result from a single cause, but from the interaction of multiple genetic, metabolic, and environmental factors. Each breach represents a failing biological mechanism that, when it lines up with others, leads to neurodevelopmental alterations.

Among these breaches, vitamin D deficiency plays a major role, compromising neuronal protection and fostering chronic inflammation. The intestinal microbiota—often imbalanced in autistic children—produces neuroactive toxins like p-cresol, which disrupt neurotransmitter regulation and worsen developmental difficulties. Added to this is immune dysregulation, in which prenatal inflammation and maternal autoantibodies alter synapse formation.

Neurotransmitter imbalances—such as excessive dopamine or poorly distributed serotonin—reinforce symptoms, particularly repetitive behaviors and social difficulties. At the same time, mitochondria—essential for cerebral energy production—are often dysfunctional, further weakening neurons. Finally, environmental exposures—such as pesticides—and chronic neuroinflammation contribute to altering brain plasticity.

This seven-breach model allows us to apprehend autism as a multifactorial disorder in which each vulnerability amplifies the others. By identifying and correcting these breaches, we can not only better understand autism, but also develop personalized interventions to lessen its impact and improve the quality of life of children and their families.

A Runaway Epidemiology: Autism in Numbers

Autism is one of the most challenging health issues of our time, not only for families but also for schools and society.

The numbers speak for themselves: the prevalence of autism has risen dramatically in just a few decades.

According to the U.S. Centers for Disease Control and Prevention (CDC), the latest estimate published in 2024 puts autism at 1 in 31 children (≈3.2%). Just a few years earlier, in 2020, the figure was 1 in 54, and back in 2000 it was 1 in 150.

In the early 1970s, it was less than 1 in 10,000. This shows we are facing not a small statistical shift, but a global wave.

The U.S. data come from the ADDM Network (Autism and Developmental Disabilities Monitoring Network), which monitors more than 300,000 children and remains the most reliable source worldwide. Yet international studies confirm the same trend: autism prevalence now ranges from 1 to 2 percent in Europe, Asia, and North America.

A Debate That No Longer Holds

Some argue the rise is only due to better awareness or broader definitions. But the evidence shows otherwise:

- Expanding the definition of autism (for example, including Asperger's in the DSM-IV) accounts for less than 10% of the increase.
- Careful follow-up studies, such as those from North Dakota and California, confirm that earlier data were methodologically sound.
- Genetic disorders do not increase like this over time. If autism were purely genetic, its prevalence should remain stable.

In 2014, Dr. Cynthia Nevison compared autism prevalence with environmental factors and concluded that at least 75–80% of the increase is real, not just due to new criteria. Similarly, Dr. Darold Treffert, who conducted one of the very first autism prevalence studies in 1970, stated in 2015 that the growth is undeniable and likely linked to environmental causes.

From Denial to Early Detection

When I was a child, autism was virtually unknown. Today, almost everyone knows a family affected by it. During my medical studies, autism was covered in less than half an hour of lectures, often under the label of "childhood psychosis." Things have changed.

Parents now come to my medical office with babies under one year old, already worried about unusual signs. Some mothers even consult me before pregnancy, asking how to reduce risks. This is a huge shift: from denial, to acceptance, to proactive prevention.

On the ground, the rise is obvious. Schools face growing demand for specialized teachers and assistants. Health systems are overwhelmed. And parents are asking for early intervention, because waiting until the age of three to diagnose autism—as was once recommended—is simply counterproductive.

A Broader Problem

The CDC also reports that 1 in 6 children between ages 3 and 17 in the U.S. has a developmental disorder, including autism, ADHD, or learning disabilities. Some analysts even predict that soon 1 in 2 children could be affected by developmental challenges.

If that happens, education systems and societies will face consequences we are not prepared for. Pretending that this is not a real epidemic is no longer an option.

A Public Health Challenge

The rise of autism is not a curiosity–it is a public health emergency. Recognizing it as such is the first step toward collective action. If we continue to see autism only through the narrow lens of genetics or through the romanticized idea of "neurodiversity," we will miss the point.

Autism must be addressed as a condition with inflammatory and environmental roots, requiring multidisciplinary care and urgent global cooperation. The children affected today will shape the adults and societies of tomorrow.

We cannot afford denial. What we are facing is nothing less than a pandemic of neurodevelopmental disorders. Accepting this

reality is the only way to mobilize resources, accelerate research, and build a future where fewer children–and families– are left without answers or support.

Beyond Diagnostic Labels: Rethinking Autism

The word *autism* first appeared in 1911, when Swiss psychiatrist Eugen Bleuler used it in his writings on schizophrenia.

In 1943, Austrian-American child psychiatrist Leo Kanner published the first scientific article describing 11 children with what he called "autistic disturbances." Later, in 1949, he introduced the term *infantile autism*.

Kanner described these children as withdrawn, living in "aloneness," and clinging to "sameness." At the time, autism was seen as an early form of childhood schizophrenia. For decades, this view dominated psychiatry.

Everything changed in 1980, when the DSM-III (Diagnostic and Statistical Manual of Mental Disorders, third edition) separated autism from schizophrenia. It introduced the category of "pervasive developmental disorders." But the criteria were too strict, and many children were left without a diagnosis. Later revisions, like the DSM-III-R (1987) and DSM-IV (1994), broadened the scope and included conditions such as Asperger's syndrome and Rett syndrome.

In 2013, the DSM-5 adopted a simpler term: *Autism Spectrum Disorder (ASD)*. It brought together autism, Asperger's, childhood disintegrative disorder, and pervasive developmental disorder not otherwise specified. In 2018, the World Health Organization followed with the ICD-11, which uses the same category. Today, *ASD* is the worldwide standard.

A Terminology Jungle

This shifting language created confusion. Parents and doctors now face a jungle of labels, especially since many conditions overlap. Autism often shares features with ADHD, learning disorders, dyspraxia, and other "dys-" conditions.

Research shows these conditions have common biological roots: oxidative stress, mitochondrial dysfunction, abnormal neurotransmitters, poor brain blood flow, and signs of inflammation. Genetics alone cannot explain the epidemic rise, because genetic disorders do not suddenly increase in prevalence.

Toward a Unified View

Some experts, like American pediatrician Michael Goldberg, suggest grouping autism and related disorders under a single "neuro-immune dysfunction syndrome." While not all agree, the idea highlights a deeper truth: these disorders are linked by immune and environmental factors, not just by genes.

From my own medical practice with thousands of children, I see that current diagnostic borders are outdated. Autism and ADHD, for example, are not completely separate. They are part of a continuum—a spectrum of neurodevelopmental conditions shaped by environment and biology.

The Role of Environment

Children today have the same genetic makeup as past generations. What has changed is the world around them: diets, chemicals, pollutants, lifestyles, and stress. These environmental shifts have created an explosion of vulnerability.

The numbers prove it. According to the U.S. Centers for Disease Control and Prevention (CDC), the latest 2024 estimate shows that 1 in 31 children (about 3.2%) has autism. In 2000 it was 1 in 150, and in the 1970s it was less than 1 in 10,000.

Even more concerning, the CDC reports that 1 in 6 children now has a developmental disorder, and some experts predict this could soon reach 1 in 2.

A Wake-Up Call

Autism is more than a set of labels or a matter of "neurodiversity." It is a sign of a larger crisis—an intersection of genes and environment producing neuro-inflammatory conditions on a massive scale.

As one striking phrase puts it, autistic children are the "prophets of modern times," warning us of the consequences of our lifestyles. Whether we agree or not, their message is clear: society must act now.

Reading Leo Kanner's First Article on Autism

The first medical description of autism dates back to 1943, when American child psychiatrist Leo Kanner published his landmark paper "Autistic Disturbances of Affective Contact." He described 11 children–8 boys and 3 girls–who showed unusual behavior compared to their peers. Two traits stood out: an extreme preference for solitude ("aloneness") and a powerful need for sameness ("sameness").

Kanner suggested a possible genetic basis because the signs appeared very early. However, he also emphasized the family environment.

All 11 children came from educated or wealthy backgrounds, and their parents were often described as emotionally distant.

This observation shaped a harmful theory that autism was caused by cold, unloving parents.

The Psychological Misinterpretation

For decades, autism was seen through a psychological lens. The idea was pushed to the extreme by Bruno Bettelheim, an Austrian-born American psychologist. In his 1967 book "The Empty Fortress," he blamed autism on "refrigerator mothers" who emotionally rejected their children.

This theory led some practitioners to remove children from their families, with tragic consequences.

Yet Kanner's original case reports already pointed to possible biological causes, especially digestive problems. Out of the 11 children he described, 9 had significant gastrointestinal issues–vomiting, constipation, feeding problems, or unusual eating behaviors.

This evidence was overlooked for more than half a century because Kanner himself was a psychiatrist, and psychology dominated the field.

Case 1: Donald Grey Triplett

Kanner's "Case 1," Donald Grey Triplett, is now recognized as the first person formally diagnosed with autism. As a child, Donald showed echolalia (repeating words), avoided other children, and ignored his mother's attempts to engage him.

Unlike many families at the time, his parents refused to institutionalize him. Thanks to his mother's persistence, Donald attended school, graduated with a college degree in French, and worked in a local bank. He lived an almost independent adult life, driving his own car and traveling the world—a future that would have been impossible had he been confined to an institution.

Early Biological Clues

Other researchers also noticed links between autism and physical health:

- In the 1950s, some studies reported autism-like behaviors in children with gluten intolerance.
- In the 1960s, physicians observed improvements when certain foods were removed from diets.
- By the 1970s, researchers found that gluten-free diets helped some autistic children with chronic diarrhea and abnormal brain activity.

Bernard Rimland's Turning Point

A major shift came with Bernard Rimland, an experimental psychologist and father of an autistic child. In 1964, he published "Infantile Autism: The Syndrome and Its Implications for a Neural Theory of Behavior," challenging the dominant psychological theories. He founded the Autism Research Institute in 1967, which gathered clinical data and opened the way to biological and biomedical research on autism.

While some of Rimland's later initiatives, like the Defeat Autism Now (DAN) program, were criticized for promoting controversial or poorly tested treatments, his main contribution was crucial: he redirected the field toward science, biology, and families' lived realities.

Lessons from Kanner's Work

Kanner's original paper revealed a puzzling condition that defied easy explanation. But his psychiatric background led him to interpret autism mainly through a psychological framework, delaying recognition of its biological complexity.

This story highlights the dangers of dogma: brilliant clinicians discovered important pieces of the puzzle but often stayed trapped in narrow theories. True understanding of autism requires moving beyond rigid categories to see it as a complex, multi-layered condition—one that bridges biology, environment, and human experience.

Stop "Psychiatrizing" Autism: It Is a Systemic Disease

In April 2019, Sophie Cluzel, France's Secretary of State for People with Disabilities, sparked controversy when she said on Radio Monte Carlo that we must "stop talking about psychiatry" when it comes to autism. She stressed the need for early detection and intervention in children, and criticized the placement of autistic adults in psychiatric hospitals, calling it "inappropriate."

Her comments triggered outrage from the psychiatric establishment. French child psychiatrists and professional associations demanded an apology, calling her remarks "outrageous, irresponsible, and scandalous." This conflict is not new. In 2011, the documentary : Le Mur (The Wall) by Sophie Robert showed psychoanalysts discussing autism. The film caused a major scandal. Some analysts sued, claiming their views had been distorted. Although a court initially banned the film, the ruling was later overturned.

Eighty Years of Psychiatry's Hold on Autism

Since the early descriptions by Leo Kanner and Hans Asperger more than 80 years ago, autism has been classified as a psychiatric disorder with no known cause, no real treatment, and little hope for prevention. Psychoanalytic theories—popularized by Bruno Bettelheim's 1967 book *The Empty Fortress*—framed autism as the result of emotionally distant "refrigerator mothers." Though discredited, this idea shaped attitudes for decades.

Today, millions of children worldwide are diagnosed with autism spectrum disorder (ASD), pervasive developmental disorder, or attention-deficit/hyperactivity disorder (ADHD).

Prevalence has reached alarming levels. In the U.S., autism affects about 1 in 31 children (CDC, 2024). These numbers continue to rise, making it increasingly hard to see autism as only a mental disorder.

The Limits of Psychiatric Diagnosis

Autism has long been defined through the psychiatric manual DSM (Diagnostic and Statistical Manual of Mental Disorders). But this framework is under fire. A psychiatric diagnosis should come only after ruling out medical, metabolic, genetic, or other causes.

Instead, many children are quickly prescribed antipsychotic drugs like risperidone, aripiprazole, or stimulants like methylphenidate. Some estimates suggest that one in ten children takes at least one of these medications.

This medicalization through psychiatry hides the truth: autism is more than behavior. It shows itself in the body. Many autistic people suffer from gastrointestinal problems, mitochondrial dysfunction, allergies, autoimmune diseases, epilepsy, folate metabolism disorders, sleep disturbances, and motor difficulties. Taken together, these comorbidities reveal autism as a systemic disease with an inflammatory component, not just a psychiatric label.

Scientific Evidence of a Systemic Condition

Research supports this systemic view:

- Studies of cytokines and immune activity show inflammation in the brain's microglia.
- New theories highlight abnormal cell danger responses and disrupted calcium signaling.
- Immune imbalance between regulatory T-cells and Th17 lymphocytes—patterns found in many systemic diseases—has also been

documented in autism.

These findings point to an inflammatory, whole-body condition. Yet the psychiatric establishment still insists autism is purely developmental and behavioral.

A Public Health Crisis

The numbers tell the story. In the 1970s, autism affected about 1 in 10,000 children. By 2000 it was 1 in 150. Today, it is 1 in 31 in the U.S. If trends continue, some analysts warn that we may approach a 1-to-1 ratio of autistic to non-autistic children within a generation. To deny this epidemic is to ignore reality.

This is more than a scientific debate–it is a public health emergency. Keeping autism inside psychiatric categories underestimates the scope of the crisis. The real picture is one of rising prevalence, enormous family suffering, and a burden on schools and health systems.

The Trap of "Neurodiversity"

In recent years, some voices have promoted the idea of autism as simple "neurodiversity," a different way of being rather than a disease. While respect and inclusion are vital, romanticizing autism as merely a difference ignores the medical realities faced by millions of children and families.

Many parents see this narrative as a barrier. Their children struggle with pain, seizures, digestive disorders, immune dysfunction, and severe developmental delays. For them, calling autism a "difference" is not comforting–it feels like denial. Pretending that biology does not matter blocks research into potentially life-changing treatments.

Breaking with Dogma

French philosopher Gaston Bachelard once wrote: "Reality is never what we could believe, but it is always what we should have thought."

Autism forces us to rethink. For too long, dogma has held back progress–whether psychoanalysis blaming parents, or rigid psychiatry reducing autism to behaviors.

We must instead see autism through the lens of modern medicine: as a systemic, inflammatory condition influenced by genes, immunity, and environment. This shift does not promise a miracle cure tomorrow. But it opens the way to earlier detection, targeted interventions, and perhaps prevention for future generations.

A Call for Change

The principle *primum non nocere*—"first, do no harm"—is ignored when children are locked into psychiatric categories and treated only with heavy drugs. Denying biology prolongs suffering.

Families deserve better. Medicine must rise to the challenge of autism as it truly is: a complex, systemic disease. Only then can we give children and their parents a real chance at relief, hope, and fulfillment.

Autism's Inflammatory Nature

Modern research strongly suggests that autism is not only a brain disorder but also an inflammatory condition affecting the whole body.

Scientists have found high levels of inflammatory molecules such as IL-1β, TNF, and IL-8 in the blood, spinal fluid, and brains of many autistic individuals.

These cytokines—chemical messengers of the immune system—are consistently abnormal in almost every study.

Other evidence supports this inflammatory theory:

- The presence of auto-antibodies against brain tissue.
- Abnormal numbers of immune cells like CD4 and CD8 lymphocytes.
- Changes in natural killer cells, growth factors, and brain proteins such as BDNF.
- Genetic mutations that disrupt immune function.

One striking finding is the role of neurotensin, a molecule found in the gut and brain. It increases inflammation by stimulating IL-1β and IL-8, and it also activates microglia, the brain's immune cells. Neurotensin may explain why allergies and food intolerances are so common in autism.

Another key player is IL-37, an anti-inflammatory molecule. It seems to counteract neurotensin's effects and has been suggested as a potential treatment target.

Studies also show an imbalance between Th17 cells (which drive inflammation) and regulatory T-cells (which calm it down). This imbalance is linked to autism severity.

Even markers like C-reactive protein (CRP), commonly used to predict heart disease, are elevated in autism and correlate with symptom severity.

All these findings point to one conclusion: autism involves chronic inflammation of the brain and body.
Recognizing this opens new doors for diagnosis, treatment, and hope for families.

Autism and Autoimmunity: A Hidden Connection

Over the last two decades, scientists have explored many factors that may contribute to autism spectrum disorder (ASD).
One of the most important areas of research is the role of the immune system. Evidence suggests that problems in the immune response, including autoimmunity, play a role in the development and severity of autism.

What Is Autoimmunity?

Autoimmunity happens when the body's defense system mistakenly attacks its own tissues. Instead of only fighting infections, the immune system produces antibodies that target the body's cells. This process is well known in diseases like type 1 diabetes, lupus, or multiple sclerosis. Many studies now suggest that similar immune problems may also occur in autism.

Immune Dysregulation in Autism

Researchers have found many immune abnormalities in children with autism:

- Higher levels of inflammatory markers (cytokines, C-reactive protein, etc.).
- Activation of microglia (immune cells in the brain) in post-mortem studies.
- Genetic differences in immune regulation.
- Abnormal responses of white blood cells like macrophages and monocytes.

These findings suggest that autism is not only a brain condition but also involves the immune system. In fact, several studies show that when immune responses are corrected—whether in lab animals or in human patients—autism-related behaviors can improve.

One 2018 pilot study by immunologist Isaac Melamed showed that treatment with intravenous immunoglobulins (IVIG) led to measurable improvements in autistic children.

Maternal Immunity and Pregnancy

Inflammation during pregnancy also appears to raise autism risk. Viral infections, autoimmunity, or high inflammation in pregnant women may influence the baby's brain. A 2014 study of 1.2 million pregnancies in Finland found that mothers with high levels of C-reactive protein (a marker of inflammation) were 43% more likely to have a child with autism.

Large population studies add weight to this link. In Denmark, researchers followed nearly 690,000 children born between 1993 and 2004.

They found that mothers with autoimmune diseases such as rheumatoid arthritis, lupus, or type 1 diabetes were more likely to give birth to autistic children.

Genes, Environment, and the HLA System

Genetics also matter. A specific gene variant, HLA-DRB1*04, has been found in both rheumatoid arthritis and autism, suggesting a

shared pathway. This points toward a "gene–environment interaction," where a genetic tendency interacts with outside factors like infections or diet. This is a common theme in many autoimmune diseases.

Autoantibodies in Autism

Autoantibodies—antibodies directed against the body itself—have been detected in many people with autism. Here are some examples:

- Anti-endothelial cell antibodies (AECA): Found at higher levels in autistic children; also seen in autoimmune diseases like lupus.
- Anti-measles antibodies: Studies found elevated levels in 83% of autistic individuals compared to controls, along with other viral antibodies.
- Anti-folate receptor alpha (FRα) antibodies: Block folate transport into the brain and may affect fetal development.
- Anti-nuclear antibodies (ANA): Found in 25% of autistic children vs 4% of controls; linked with severity.
- Anti-ribosomal P protein antibodies: Found in 58% of autistic children vs 5% of controls; can attack nerve tissue.
- Anti-gliadin antibodies: Found in gluten-related conditions and autism; may trigger immune reactions in the brain.

Family History of Autoimmunity

Autoimmune diseases also appear more often in families of autistic children. Conditions like type 1 diabetes, inflammatory bowel disease, and rheumatoid arthritis occur at higher rates in these families, suggesting a shared vulnerability.

Neurodevelopmental Disorders With Autoimmune Features

Some related conditions show even clearer links to autoimmunity:

- PANS (Pediatric Acute-onset Neuropsychiatric Syndrome).
- PANDAS (Pediatric Autoimmune Neuropsychiatric Disorders Associated with Streptococcus).

These disorders share features with autism and are clearly autoimmune in nature. Treatments like anti-inflammatory drugs, steroids, high-dose vitamins, IVIG, and even plasmapheresis sometimes help these children.

Potential Treatments and Controversies

Because of these findings, immune-based treatments are being studied. IVIG has shown promise, though it remains expensive and experimental. Other options—like anti-inflammatory drugs or biologic therapies—are still being tested. The challenge is that autism is highly diverse: not all children will respond to the same interventions, and careful research is needed.

A Shift in Perspective

For decades, autism was treated mainly as a psychiatric or developmental disorder. The possibility of an autoimmune basis challenges

this view. It suggests that at least some cases of autism might respond to medical treatments aimed at correcting immune dysfunction, rather than only behavioral therapy or psychiatric drugs.

This does not mean that autism is "just an autoimmune disease." But it does mean that the immune system may play a major role in how autism develops and how severe it becomes. Understanding this link may lead to new therapies, and more importantly, new hope for families.

Autism and Mast Cell Activation

Mast cells are a type of white blood cell that store many chemicals in tiny granules, such as histamine, serotonin, and tryptase.

When they are triggered—usually by allergens—they "degranulate," releasing these chemicals and causing allergic reactions.

But mast cells can also release other substances like cytokines (IL-4, IL-6, IL-1, TNF), prostaglandins, and chemokines.

These molecules don't just create allergy symptoms—they can also fuel inflammation throughout the body and brain.

Many studies show that allergic diseases in children are strongly linked with psychological or behavioral problems.

Chronic allergic reactions can create long-term inflammation, which affects the nervous system.

For children with autism, this ongoing inflammation may worsen neurological and behavioral symptoms.

In fact, many autistic children show food intolerances or allergy-like symptoms even when blood tests are negative.

This suggests mast cells may be activated by non-allergic triggers such as stress, toxins, or gut imbalances.

One large study of over 92,000 children found strong statistical links between eczema, ADHD, and autism.

Another study confirmed higher rates of asthma, psoriasis, and autoimmune diseases in autistic children.

Children with mastocytosis—a rare disease with excess mast cells—have up to ten times the risk of autism compared to other children.

Traditionally, mast cells were thought to live mostly in the skin, gut, and near blood vessels.

But newer research shows they also exist in key brain areas like the thalamus and hypothalamus, which control emotions, stress, and body regulation.

When mast cells in the brain are activated, they can disrupt the blood-brain barrier, allowing immune cells to invade the brain and triggering microglial activation.

This process looks like a small, localized brain inflammation (encephalitis).

Two important triggers are CRH (corticotropin-releasing hormone)–released during stress–and neurotensin, a molecule found at high levels in autistic children.

Neurotensin is concentrated in brain regions linked to language and emotion, making its role even more significant.

Together, CRH and neurotensin over-activate mast cells, increasing brain inflammation.

Mast cells also release mitochondrial DNA (mtDNA) and ATP when activated.

Normally, the body sees these molecules as "foreign," because mitochondria evolved from ancient bacteria.

Their release fuels more inflammation, creating a vicious cycle that can damage neurons.

Elevated mtDNA has been found in the blood of autistic children.

Mast cells also interact with Th17 immune cells, which are already known to play a role in autism-related inflammation.

By releasing IL-6, TNF, and other cytokines, mast cells push the immune system toward a chronic pro-inflammatory state.

The work of Dr. Theoharides at Tufts University highlights mast cells as the "immune gateway to the brain."

He suggests that at least a subgroup of autism may be understood as a form of "brain allergy."

Overall, mast cells appear central in linking allergies, stress, and brain inflammation in autism.

Understanding and targeting mast cell activation may lead to new, personalized therapies–especially for autistic children with allergies or food sensitivities.

By calming mast cell activity, it may be possible to reduce neuroinflammation, improve symptoms, and offer families more effective and lasting treatment options.

Autistic Regression: A Controversial Concept

One of the most debated topics in autism research is autistic regression.

When Dr. Leo Kanner first described autism in 1943, he did not mention regression. He suggested autism started at birth or very soon after. But by the 1960s, some doctors reported that children with autism could actually lose skills they had already gained. In the 1980s, more studies confirmed this idea, especially focusing on the loss of early language skills. Since then, many scientists and clinicians have recognized regression in at least a subgroup of autistic children, usually in the first years of life.

Regression means that parents report their child seemed to develop normally for the first year or two, and then lost abilities–often language, social interaction, or motor skills. This loss can be sudden or gradual.

A large meta-analysis of 85 studies estimated that 32% of autistic children experience regression, with the average age of onset at 21 months. However, real numbers may be higher. Definitions vary from study to study, and in clinical practice regression is seen more often than statistics suggest.

Does Regression Change Prognosis?

Whether regression leads to worse long-term outcomes is controversial. Some studies show no difference between regressive and non-regressive autism. Others suggest children with regression may struggle more with language, adaptive functions, behavior, and social skills.

Regression is rare in most other neurodevelopmental disorders. Exceptions include the Landau-Kleffner syndrome (childhood epilepsy with language loss), Rett syndrome (a genetic disorder in girls), and certain metabolic diseases such as lysosomal storage disorders. Childhood disintegrative disorder, marked by regression around age 3, is another rare example.

Searching for Causes

Many studies have tried to find biological triggers for regression. Vaccines, gastrointestinal issues, mitochondrial dysfunction, and epilepsy have all been investigated, but no clear cause has been confirmed.

Dr. Richard Kelley, a specialist at Johns Hopkins, offered one of the most interesting theories. Based on years of clinical work at the Kennedy Krieger Institute, he proposed that regression often results from mitochondrial dysfunction—specifically a problem in Complex I of the mitochondrial respiratory chain.

Mitochondria are the "power plants" of the cell. If their function is weakened, even slightly, the brain may not get enough energy. According to Dr. Kelley:

- Most children who regress show subtle mitochondrial problems, like mild low muscle tone.
- Regression is often triggered by infections (ear infections, strep throat, viral illnesses) or, less commonly, after vaccination.
- The vulnerable window is 12 to 30 months of age.
- Most children regress only once, though some experience multiple episodes.
- Partial recovery often occurs between ages 3 and 10.

Dr. Kelley explained that if a child already has a 50% reduction in Complex I activity, another small drop (5–10%) caused by illness or inflammation can push the system into cell death (apoptosis).

Role of Inflammation and Diet

Dr. Kelley also highlighted the role of inflammation in destabilizing mitochondria. Modern diets, with high omega-6 fats and lower omega-3 intake, may increase inflammation and worsen vulnerability.

He also pointed to the rise in acetaminophen (paracetamol) use since the 1980s as a possible contributor. Acetaminophen metabolism produces NAPQI, a toxic byproduct that damages mitochondrial membranes and reduces energy production. This could help explain why regression rates increased as aspirin use declined and acetaminophen became the standard fever reducer for children.

Beyond the Regression vs. Non-Regression Divide

A 2024 study in the International Journal of Molecular Sciences compared autistic children with and without regression. It suggested that:

- In "non-regressive" autism, immune dysfunction and inflammation may start first, and mitochondrial problems appear later.
- In "regressive" autism, the initial problem is mitochondrial, and immune dysfunction comes second.

This means regression may not be a completely separate type of autism, but rather a different entry point into the same cycle of inflammation and mitochondrial stress.

The Bigger Picture

Regression should be seen as part of a broader inflammatory model of autism. The real "point of no return" may involve what scientists call the Cell Danger Response (CDR)—a state where cells stay "on alert" after an immune or toxic insult.

Instead of debating whether regression "exists," researchers should study why and how it happens, and how to intervene early. Brain plasticity means that timely treatment and prevention could make a major difference.

Conclusion

Autistic regression is real, common, and biologically meaningful. It highlights the fragile balance between mitochondrial energy production, immune regulation, and environmental triggers. While not all autistic children regress, the fact that many do suggests that autism should not be seen as fixed or static.

Understanding regression not only helps families make sense of what they see, but also pushes science toward targeted treatments–reducing inflammation, supporting mitochondria, and adapting diets.

Autism, regression included, is best viewed as part of a systemic disorder with inflammatory and metabolic roots. By breaking away from outdated debates and focusing on biology, medicine can offer children not just management, but real hope.

Autism as a Mitochondrial Disease

Mitochondria are small structures inside almost all human cells. They look like tiny power plants, producing the energy our bodies need in the form of ATP (adenosine triphosphate). While red blood cells have none, muscle and nerve cells can have thousands of them, showing how important they are for organs that need a lot of energy.

Besides producing energy, mitochondria help regulate calcium, control cell death, generate free radicals, and even play a role in immune and antiviral responses. Because of these many tasks, when mitochondria do not work well, many systems in the body can be affected.

The brain is especially dependent on mitochondria. It consumes nearly 20% of the body's energy, even though it makes up only 2% of body weight. Developing neurons in infants and young children are very sensitive to mitochondrial problems. This is why early mitochondrial dysfunction can strongly affect brain growth and development.

When mitochondria produce energy, they also release free radicals–unstable molecules that can damage proteins, lipids, and DNA. Normally, antioxidant enzymes like catalase and superoxide dismutase neutralize them. But if free radicals build up, cells undergo oxidative stress, which can lead to damage and death. The brain is especially vulnerable to this because it has high energy needs, a high fat content, and weaker antioxidant defenses than other organs.

Mitochondria and Brain Disorders

Research has linked mitochondrial dysfunction not only to aging but also to brain diseases like Parkinson's, Alzheimer's,

Huntington's disease, and schizophrenia. Many metabolic and heart diseases are also connected to mitochondria.

In the 1980s, doctors first noticed that some children with autism had high blood lactate levels, a possible sign of mitochondrial problems. In 1998, J. Lombard proposed the hypothesis that autism could be considered a mitochondrial disease. Evidence included high lactate, unusual Krebs cycle metabolites in urine, low carnitine, and poor brain use of glucose and ATP.

Since then, studies have shown many mitochondrial abnormalities in autistic children. These were found in blood, cerebrospinal fluid, urine, muscle biopsies, fibroblasts, and even brain tissue. The most common issues are problems with Complex I of the electron transport chain, sometimes combined with defects in Complex III and IV.

Key Studies

A 2010 study published in JAMA (University of California, Davis) found that about 80% of children with autism showed biological markers of mitochondrial dysfunction. Compared with control children, they had:

- Lower Complex I activity
- Higher levels of pyruvate and lactate in blood
- Lower activity of pyruvate dehydrogenase
- More hydrogen peroxide (a damaging free radical)

Another study of post-mortem autistic brains showed reduced function of mitochondrial complexes in the frontal and temporal lobes and in the cerebellum. This suggested defective energy metabolism and more oxidative stress. Interestingly, these deficits seemed less obvious in adult brains, hinting at possible adaptation over time.

Later research (2013) confirmed reduced activity of Complex I and IV, less antioxidant enzyme SOD2, and more oxidative DNA damage. It also showed abnormal mitochondrial "dynamics": autistic brains had more proteins linked to fission (splitting) and fewer linked to fusion (joining), meaning their mitochondria were more fragmented and unstable.

In 2014, Dr. Suzanne Goh (Columbia University) used magnetic resonance spectroscopy to measure lactate in the brain. Autistic patients had higher levels, providing the first in vivo imaging evidence of mitochondrial dysfunction.

Symptoms Connecting Autism and Mitochondria

Many clinical signs of mitochondrial disease overlap with autism:

- Seizures (epilepsy)
- Gastrointestinal problems (constipation, reflux, diarrhea)
- Motor difficulties, low muscle tone, muscle pain
- Growth problems
- Sensory difficulties
- Learning delays and developmental regression
- Diabetes and thyroid issues
- Migraines, breathing problems, lactic acidosis

This overlap supports the idea that at least a subgroup of autism is linked to mitochondrial dysfunction.

Environmental Triggers

Mitochondrial dysfunction can be primary (genetic) or secondary (caused by external factors). Secondary dysfunction is common in autism. Triggers include:

- Environmental toxins (pesticides, PCBs, heavy metals)
- Medications (valproate, aspirin, acetaminophen/ paracetamol)
- Short-chain fatty acids from gut bacteria, especially propionic acid
- High tumor necrosis factor (TNF-alpha)
- Excess nitric oxide
- Folate deficiency
- Low glutathione synthase
- Oxidative stress

These factors either reduce ATP production directly or increase oxidative damage, stressing mitochondria further. If a child already has a genetic vulnerability, the damage can be greater.

Immune System and Inflammation

Mitochondria are closely linked to the immune system. In autistic children, autoimmune diseases are more common than in the general population. Immune cells themselves show mitochondrial deficits.

Inflammation consumes energy. Cytokines (inflammatory molecules) force cells to burn more ATP, leading to exhaustion of energy supplies. This means that inflammation can cause mitochondrial problems, and mitochondrial problems can worsen inflammation—a vicious cycle.

This also explains why autistic regression (loss of language or social skills after early normal development) often follows infections. Infections may trigger immune responses that destabilize already fragile mitochondria.

Sex Differences

Autism affects boys more than girls (about 4.5 to 1). Animal studies show that male brains may produce more reactive oxygen species and be more vulnerable to mitochondrial defects. This may partly explain the higher prevalence in males.

Therapeutic Approaches

If autism has a mitochondrial basis, improving mitochondrial function could help symptoms. Several supplements have been studied:

- L-carnitine (helps transport fatty acids into mitochondria; clinical trials showed improvements in behavior and cognition)
- Coenzyme Q10 / Ubiquinol (supports electron transport and reduces oxidative stress)
- B vitamins (e.g., methylcobalamin), vitamin C, and vitamin E

- N-acetyl-cysteine (NAC), a glutathione precursor and antioxidant
- Carnosine, sulforaphane, luteolin (antioxidants with promising results)

A randomized controlled trial of multivitamins plus mitochondrial cofactors showed improvements in glutathione, ATP, oxidative stress markers, and also in autism-related behaviors (hyperactivity, stereotypies, social interaction, receptive language). These therapies are not cures, but they support the idea that mitochondrial dysfunction is treatable and may improve quality of life.

The Cell Danger Response

Dr. Robert Naviaux has proposed the Cell Danger Response (CDR) theory. Mitochondria detect toxins or infections and trigger protective shutdowns to help the body heal. But if this system is activated too often, it prevents normal recovery and leads to chronic disease, including autism. This view suggests that autism is not just a "brain wiring problem" but a metabolic and systemic condition that starts in the mitochondria.

Conclusion

Autism has long been seen only as a psychiatric or developmental disorder. However, growing research shows that mitochondrial dysfunction may play a central role in many cases. From energy failure and oxidative stress to immune dysregulation and regression, mitochondria may be the missing link connecting many symptoms.

If we see autism partly as a mitochondrial disease, new strategies open up: preventing exposure to toxins, identifying children at risk, using targeted supplements, and treating inflammation early. This perspective does not deny the complexity of autism–it adds a biological foundation that could guide better prevention, diagnosis, and therapy. By focusing on mitochondria, we may be able to reduce suffering and improve the lives of children and families facing autism.

The Genetic Puzzle of Autism

For decades, scientists have been searching for a clear genetic cause of autism. Since the 1970s, researchers have looked at the human genome to see if there was a single "autism gene." Thanks to modern DNA sequencing, this search has expanded, but the results have been more confusing than clear.

No Single Autism Gene

There is no one gene that explains autism. Instead, many different genetic changes can contribute. For example, Fragile X syndrome is linked to a mutation in the FMR1 gene, and Rett syndrome is caused by mutations in the MECP2 gene. Both can lead to autistic features. So far, scientists have found around 65 genes strongly linked to autism and over 200 more with weaker links. These genes often affect how brain cells communicate or how other genes are turned on and off.

Small Variations, Big Effects

Each human has thousands of small genetic differences, called single nucleotide polymorphisms (SNPs). Some SNPs affect how our bodies use nutrients, fight infections, or react to medicines. Certain SNPs, such as those in COMT, MTHFR, and other genes, may increase autism risk, especially when combined with environmental stress.

Another type of genetic change is called a copy number variation (CNV), where chunks of DNA are missing or repeated. Many CNVs have been reported in people with autism, but they vary widely and are not the same in everyone.

Large Research Projects

Big projects like MSSNG (funded by Google) and SPARK (funded by the Simons Foundation) are sequencing the DNA of thousands of autistic people and their families. The goal is to build a genetic map that helps researchers understand the many forms of autism. A Chinese database called AutismKB has already listed more than 3,000 genes linked to autism.

Beyond the Nuclear Genome

Genes are only part of the story. Most of our DNA does not code for proteins–it was once called "junk DNA," but now scientists think it plays important roles in gene regulation. The mitochondria, which are the "power plants" of cells, also have their own DNA, passed down from the mother. Studies have shown that children with autism sometimes have more deletions in mitochondrial DNA, which could affect how their brains use energy.

The Role of the Microbiome

The human microbiome–the collection of bacteria in our gut–also carries millions of genes. Research suggests that these microbial genes interact with human genes and may influence brain development and behavior. For example, certain gut bacteria produce chemicals that affect the brain.

Genetics vs. Environment

A large study of over 2 million children in five countries found that about 81% of autism risk is linked to inherited genetic factors. But this does not mean that genes alone determine autism. Rapid increases in autism rates cannot be explained by genetics alone. Environmental factors, immune responses, and epigenetics (chemical changes that turn genes on or off) are also important. For example, inflammation, toxins, or nutrition can all interact with genes to influence brain development.

The Bigger Picture

Genetics is an important piece of the autism puzzle, but it is not the whole story. Autism likely results from a combination of genetic predisposition, environmental influences, and biological processes like immune activity and mitochondrial function. Rather than searching for a single "autism gene," science is moving toward understanding how these different factors interact. This broader approach may open the door to better prevention and treatment strategies in the future.

Autism and Genetic Variations: Understanding SNPs

Autism Spectrum Disorder (ASD) is a complex condition influenced by many biological and environmental factors.

For decades, scientists have tried to find genetic explanations for autism. One promising area of research involves single nucleotide polymorphisms (SNPs). SNPs are small changes in DNA – a switch of a single "letter" in the genetic code. While tiny, these changes can sometimes affect how the body works.

What are SNPs?

Every person's DNA is made of about 3.2 billion "letters" (A, T, C, and G). Sometimes, one of these letters is swapped for another. This is called a SNP. Most SNPs are harmless and simply contribute to the natural diversity between people. For example, some SNPs affect eye color, hair type, or even how we respond to certain medications.

However, some SNPs can influence health and increase the risk of certain diseases.

SNPs and Autism

Research shows that people with autism often carry a higher number of certain SNPs compared to the general population.

These variations do not guarantee autism, but they may increase susceptibility. Scientists now think that autism does not come from a single "autism gene." Instead, it is the result of many genetic factors working together, often combined with environmental triggers.

Some SNPs affect genes involved in brain communication. Neurons (nerve cells) talk to each other using chemical messengers called neurotransmitters. Genes related to serotonin, dopamine, and GABA – key brain chemicals – are often influenced by SNPs linked to autism. If these signals are disrupted, brain circuits may develop differently, which could affect language, social skills, and behavior.

Other SNPs affect genes related to the immune system, energy production in mitochondria, and even how the body handles toxins. This shows that autism is not only about the brain; it involves multiple systems in the body.

Genetics vs. Environment

One important lesson from SNP research is that genes are not destiny. Having a SNP does not automatically mean someone will develop autism. Instead, genes create a vulnerability. Environmental factors – such as infections during pregnancy, nutritional deficiencies, or exposure to certain chemicals – may "activate" this vulnerability. This is where epigenetics comes in. Epigenetics refers to changes that influence whether a gene is turned "on" or "off," without altering the DNA itself. Stress, diet, and environment all play roles in epigenetic regulation.

The Role of Mitochondria

Mitochondria are the "powerhouses" of the cell, producing energy for the brain and body. Because the brain consumes so much energy, it is especially vulnerable to mitochondrial problems. Some SNPs are directly linked to mitochondrial function. If mitochondria cannot produce enough energy, cells may become stressed and even die. Research shows that many children with autism have signs of mitochondrial dysfunction, which could explain why symptoms often appear after stressors such as infections.

Mitochondrial DNA and Beyond

Most of our DNA is stored in the nucleus of the cell. But mitochondria have their own DNA, passed down only from the mother. Research shows that deletions or mutations in mitochondrial DNA are more common in autism. This suggests that both nuclear DNA and mitochondrial DNA interact in complex ways to shape brain development.

Another exciting field is the study of the microbiome – the collection of trillions of bacteria living in the human gut.
These microbes contain millions of genes, many more than the human genome. Scientists believe that interactions between human genes, microbial genes, and the environment may strongly influence autism.

Why SNP Research Matters

Studying SNPs may help us develop more personalized approaches to autism. For example, in the future, genetic testing could identify which biological pathways are most affected in a child. This could guide targeted interventions – whether medical, nutritional, or behavioral. While we are not there yet, SNP research is laying the foundation for personalized medicine in autism.

Challenges

Despite progress, SNP research faces challenges.

- Complexity: Thousands of SNPs may contribute to autism, but each one has a small effect.
- Variability: Different studies report different results, making it hard to draw firm conclusions.
- Ethics: Genetic testing raises questions about privacy, access, and how results are used.

Moving Forward

Instead of looking for "the autism gene," researchers are now mapping the networks of genes and pathways involved.
The future likely lies in combining genetics with epigenetics, environmental research, and neuroscience. By integrating all these areas, science can move toward prevention, earlier detection, and better interventions.

Conclusion

Autism is not caused by a single factor. It is a complex condition shaped by many genetic variations, environmental influences, and biological systems working together. SNP research has shown that tiny changes in DNA can have wide-reaching effects on brain development and function. While we are far from complete answers, these discoveries bring hope. They remind us that autism is not one disorder but many pathways leading to similar outcomes. Understanding these pathways is the first step toward more personalized and effective care for individuals and families living with autism.

Genetic Polymorphisms and Autism: The Case of MTHFR

One of the most debated examples of how genes may be linked to autism involves changes, or polymorphisms, in the MTHFR gene. MTHFR stands for methylenetetrahydrofolate reductase, an enzyme that plays a key role in how the body uses folate (a B vitamin) and processes homocysteine, an amino acid. These processes are essential for DNA methylation, which is the way cells "turn genes on or off."

The MTHFR enzyme changes a molecule called 5,10-methylenetetrahydrofolate into another form, 5-methyltetrahydrofolate (5-MTHF). This product then works with vitamin B12 to convert homocysteine into methionine. Methionine is important not only for making proteins but also because it creates S-adenosylmethionine (SAMe), a compound essential for DNA methylation. Homocysteine can also be recycled into cysteine and glutathione, which protect cells against damage.

Because methylation is central to gene expression, many scientists believe that defects in this pathway could contribute to autism. In fact, research has focused on two common MTHFR polymorphisms, called C677T and A1298C. These variants lower the activity of the enzyme. For example, people with two copies of the C677T variant may lose up to 70% of normal enzyme activity. This can lead to low levels of folate, high homocysteine, and potential problems with neurotransmitters like serotonin and dopamine.

A meta-analysis in 2013, which reviewed eight studies including 1,672 people with autism and 6,760 controls, found a clear association between the C677T variant and higher autism risk. Still, having this genetic variant does not mean someone will definitely develop autism. Genes increase risk, but they rarely act alone.

The prevalence of these polymorphisms varies worldwide. Among people of European descent, 6–14% may carry the C677T variant in both copies of their gene. In African populations, the rate is closer to 2%. A study in Morocco found a rate of about 6%. These numbers suggest that millions of people carry the variant, but only a small subset develop autism. This shows the importance of interactions between genes and environment.

Environmental factors may strongly influence how these polymorphisms play out. Nutrition is one key factor. Most pregnant women today take folic acid supplements, but folic acid is a synthetic form that needs to be converted into methylfolate, the active version. For women with MTHFR variants, this conversion is less efficient. Some researchers argue that using methylfolate directly could be safer and more effective, ensuring the fetus gets the folate it needs for brain development. Other environmental factors—like vitamin B deficiencies, amino acid imbalances, heavy metals, or endocrine disruptors—may also interact with MTHFR variants.

The big lesson is that autism is not caused by a single gene. Instead, autism seems to come from a mix of genetic vulnerabilities, environmental exposures, and epigenetic changes (changes in how genes are expressed without changing DNA itself). The MTHFR story is a good example of this complexity. It shows how a genetic variant can set the stage, but outside factors often determine the outcome.

Understanding polymorphisms like MTHFR may help guide personalized medicine in the future. For example, knowing a child's genetic profile could help doctors recommend better forms of vitamins or nutrients, or design more targeted interventions. While much more research is needed, these insights already challenge some current practices. For example, instead of recommending standard folic acid for everyone, some experts suggest considering methylfolate, the active form, especially for women with MTHFR variants.

In summary, MTHFR polymorphisms do not "cause" autism on their own, but they highlight how genes and environment work together. They also show the importance of tailoring health strategies to each person's biology, a step that could improve prevention and care for children at risk.

Autism and the So-Called "Junk DNA"

For many years, scientists believed that only a tiny fraction of our DNA truly mattered. About two percent of the human genome contains the instructions for building proteins, the molecules that keep us alive. The other 98% was dismissed as "junk DNA," considered useless leftovers from evolution.

But the story of autism, genetics, and environment shows us how wrong that assumption was.

From Trash to Treasure

What used to be called "junk" is now seen as a powerful regulator of our genes. This non-coding DNA helps decide **when, where, and how** genes are expressed. Instead of being silent, it works like a conductor of an orchestra, telling different genes when to play.

It contains switches called promoters, enhancers, and silencers that can turn genes on or off. It also holds repeated sequences, transposons, and pseudogenes that shape the stability of the genome. Far from being useless, these elements influence how cells grow, develop, and adapt.

The Princeton Breakthrough

A few years ago, a team led by Olga Troyanskaya at Princeton University published a groundbreaking study in *Nature Genetics*. Using artificial intelligence, they analyzed the genomes of nearly 1,800 families with an autistic child. None of the other family members had autism.

What they discovered was striking: children with autism carried an unusually high number of mutations in the non-coding regions of their DNA. These mutations were not inherited. They appeared spontaneously, setting these children apart from their relatives.

It was the first clear evidence that non-coding DNA mutations could be directly linked to autism.

Why It Matters

If 98% of our DNA plays a role in regulating the rest, then ignoring it was like trying to understand a symphony by listening to only two instruments. Non-coding DNA decides which genes are active and when. A small mutation in one of these control switches can disturb the delicate balance of brain development.

Not every mutation in non-coding DNA will cause disease. But when a mutation disrupts the proteins that control brain functions—especially those already linked to autism—the impact can be profound.

Beyond the Myth of a "Single Gene"

For decades, researchers searched for "the autism gene." Today, that idea feels naïve. Autism is not caused by a single gene but by a complex network of genetic, environmental, and epigenetic interactions.

The discovery of non-coding mutations deepens this complexity. Instead of looking for one culprit, scientists must map the three-dimensional structure of the genome and explore how hidden regulators shape gene activity.

In other words, the "junk" once dismissed may be one of the biggest keys to understanding autism.

A New Paradigm

This change in perspective mirrors a broader truth about science: what we dismiss today might turn out to be tomorrow's breakthrough. Just as "junk DNA" has proven essential, our understanding of autism is evolving.

The path forward will likely depend on combining genetics, epigenetics, and environmental science. Machine learning and other computational tools will help decode these interactions, revealing how small changes in non-coding DNA can set off larger cascades in brain development.

Looking Ahead

Understanding non-coding DNA won't only transform autism research. It may also open new doors for cancer, cardiovascular disease, and autoimmune disorders.

For autism, this knowledge could lead to personalized therapies, targeting the regulatory regions of DNA that disrupt brain function. It could also help explain why autism's prevalence is rising so sharply—something a single "autism gene" could never explain.

As I reflect on these discoveries, one lesson stands out: there is no such thing as useless DNA. Every piece of the genome has a role to play. And within those hidden regions may lie not only the secrets of autism, but also the future of medicine itself.

Autism and the Forgotten Organ: The Gut Microbiome

For a long time, medicine ignored the hidden world inside the human gut. More than 100 trillion microorganisms live there, belonging to at least a thousand different species. Today, this community of microbes, called the gut microbiome, is recognized as essential to health. Scientists even call it a forgotten organ, because it works like a complex system that protects, nourishes, and shapes us in ways we are only beginning to understand.

The story of autism and the gut begins at birth. A child born through natural delivery receives the mother's vaginal microbes, starting life with a diverse microbial community. Babies born by cesarean section miss this transfer, which may have long-term consequences. From then on, diet, antibiotics, infections, stress, and other factors continue to shape the microbiome. In return, this microbial world influences how our genes are expressed, how our immune system develops, and even how our brains function.

Half of the body's immune cells live in the gut, in constant dialogue with microbes. The intestinal wall and the immune defenses form a double barrier that protects us from harmful invaders. Friendly bacteria help maintain this barrier, produce short-chain fatty acids that feed intestinal cells, and prevent the overgrowth of pathogens. They also regulate intestinal movement, assist in absorbing vitamins and minerals, and participate in metabolic processes. Over the past two decades, evidence has grown that this internal ecosystem communicates directly with the brain. Scientists now talk about the gut-brain axis: a two-way communication system between microbes and mind.

How does this work? The mechanisms are varied. Microbes can release signaling molecules such as cytokines into the blood. They influence the vagus nerve, which links the gut to the brain. They affect tryptophan metabolism, a key pathway for producing serotonin and other neurotransmitters. Certain gut bacteria even produce neurotransmitters themselves. Lactobacillus and Bifidobacterium make GABA, while other species produce noradrenaline or serotonin. It is no surprise, then, that disturbances in the microbiome could have neurological effects.

Children with autism often suffer from gastrointestinal problems such as diarrhea, abdominal pain, or constipation. A meta-analysis found that autistic children are three times more likely to have such symptoms than neurotypical children. This observation raised an important question: could the microbiome play a role in both the digestive and neurological aspects of autism?

Over the years, researchers have noticed differences in the microbiomes of autistic children. Some studies found reduced microbial diversity. Others reported higher or lower levels of certain bacteria, including Clostridium, Desulfovibrio, Lactobacillus, and Sutterella. These imbalances, called dysbiosis, may weaken the intestinal barrier. A leaky gut could allow bacterial toxins and metabolites to escape into the bloodstream, reaching the brain and triggering inflammation.

One striking line of research focuses on short-chain fatty acids. These molecules, produced by bacterial fermentation of dietary carbohydrates, act as chemical messengers. Acetic, propionic, and butyric acids are the most common. Dr. Derrick MacFabe in Canada showed that injecting propionic acid into the brains of rodents produced behaviors and brain changes similar to those seen in autism. He suggested that propionic acid, produced by certain gut bacteria and used as a food preservative, might alter mitochondrial function and gene regulation in ways that influence brain development.

The link between gut and autism is not just an abstract scientific idea–it has a human story. In the late 1990s, Ellen Bolte, a computer scientist and mother, noticed her young son regressed into autism after multiple rounds of antibiotics for ear infections. She believed the antibiotics had destroyed his healthy gut bacteria, allowing harmful bacteria to flourish and release toxins. Her hypothesis, published in a medical journal, was dismissed by most doctors. Yet one pediatric gastroenterologist, Richard Sandler, listened. He prescribed vancomycin, an antibiotic usually reserved for severe intestinal infections. For eight weeks, Ellen's son improved dramatically, only to regress again when the treatment stopped. Dr. Sandler published these results in 2000, showing temporary but notable improvements in autistic behaviors. The study hinted that certain bacteria, such as Clostridium, might be involved.

Her story inspired other researchers, including Sydney Finegold, an infectious disease specialist who demonstrated that bacterial toxins can influence the brain through the vagus nerve. Still, for many years, funding and credibility went mostly to genetic studies, leaving gut research in the shadows. Small sample sizes, inconsistent methods, and the difficulty of obtaining mucosal samples instead of stool samples made results hard to compare. Yet the evidence continued to accumulate.

One major obstacle was diet. The microbiome changes quickly in response to food—sometimes in just 24 hours. A Western diet, high in fat and animal protein but low in fiber, reduces beneficial bacteria like bifidobacteria. In contrast, a Mediterranean diet increases good bacteria such as lactobacilli and lowers harmful ones like certain clostridia. Families of autistic children often try restrictive diets—gluten-free, casein-free, or specific carbohydrate regimens—making it even harder to standardize research. Still, diet clearly matters.

Perhaps the most convincing evidence came in 2019 from a study led by Sarkis Mazmanian at the California Institute of Technology. His team transplanted gut microbes from autistic children into germ-free mice. The mice developed autistic-like behaviors: less social interaction, repetitive actions, and slower movements. Their offspring also showed altered microbial profiles. Analysis revealed differences in dozens of metabolites. Two in particular—taurine and 5-aminovaleric acid—were low. Supplementing these metabolites improved social behaviors in the mice, suggesting a direct microbial influence on the brain.

Other studies have pointed to bacterial metabolites such as p-cresol. Found in higher levels in the urine and stool of autistic children, p-cresol and its derivative p-cresyl sulfate can affect social behavior in animals. How exactly these compounds act in the brain is still unclear, but their presence highlights the importance of microbe-derived chemicals in neurological development.

Research has also expanded to include fungi, viruses, and archaea in the gut. A large study from Hong Kong analyzed thousands of microbial genomes and identified specific multikingdom markers that distinguished autistic from non-autistic children. These included microbial pathways related to essential nutrients such as thiamine and ubiquinol. Such findings suggest that the microbiome could one day be used not only to understand autism but also to help diagnose it.

Closer to home, researchers have observed how early life events shape the microbiome of children. Lack of breastfeeding, frequent antibiotic use, cesarean birth, infections, and poor diet all reduce beneficial microbial diversity. Exposure to pesticides, food additives, and even parental stress also play a role. These factors combine with genetic and immune vulnerabilities, creating a complex web that may increase the risk of autism.

Today, the microbiome is no longer seen as a passive collection of microbes. It is an active organ that influences immunity, metabolism, and brain function. While no single bacterium or metabolite can be blamed for autism, the evidence suggests that imbalances in this microbial world contribute to symptoms. Understanding this connection may lead to new therapies: targeted probiotics, dietary interventions, or even microbiota transplants. What once seemed far-fetched is now a serious field of research.

The forgotten organ is finally being remembered. As scientists continue to unravel its mysteries, the story of the gut microbiome may become a turning point in how we understand and treat autism. What began with the observations of a determined mother has grown into an international scientific effort. The gut, long overlooked, may hold keys not only to autism but to a deeper understanding of human health itself.

p-Cresol and Propionic Acid: The Gut-Brain Connection in Autism

For years, scientists have suspected that the gut and the brain are deeply connected, and research on autism has brought this idea into sharper focus. Two molecules in particular—p-cresol and propionic acid—have captured the attention of researchers worldwide.

p-Cresol, also known as 4-methylphenol, is a toxic byproduct made when certain gut bacteria break down the amino acid tyrosine. Studies show that many autistic children have unusually high levels of p-cresol in their urine and stool. This appears linked to leaky gut, a condition where the intestinal barrier becomes more permeable, allowing bacterial products to escape into the bloodstream. High urinary levels of p-cresol have even been correlated with more severe autism symptoms.

Animal studies have provided striking evidence. In one experiment, autistic-like mice exposed to p-cresol developed social avoidance, repetitive behaviors, and abnormal dopamine activity in brain regions such as the amygdala and striatum. Even their brain wiring was affected: researchers observed poor myelination in the prefrontal cortex. In cell studies, p-cresol disrupted the normal development of oligodendrocytes, the cells that form the brain's insulating myelin sheath.

So far, scientists have identified over 55 bacterial strains capable of producing p-cresol. They belong to diverse families, including Clostridiaceae and Enterobacteriaceae. p-Cresol also promotes the growth of Clostridioides difficile, a bacterium often found in abundance in autistic patients, possibly worsening symptoms.

Another compound, hydroxyphenylpropionic acid (HPHPA)—also made by Clostridium—has been linked to autism. Both p-cresol and HPHPA inhibit an enzyme called dopamine-beta-hydroxylase (DBH). When DBH is blocked, dopamine builds up to toxic levels. Excess dopamine can damage mitochondria, the energy factories of brain cells, and create harmful oxygen radicals.

Indeed, many autistic children have elevated dopamine and homovanillic acid (HVA) levels in urine and cerebrospinal fluid. Interestingly, treatment with vitamin B6 often reduces these abnormalities and improves symptoms, suggesting a biochemical link.

HPHPA also depletes coenzyme A (CoA), a molecule essential for energy production, lipid synthesis, and brain development. Scientists believe CoA shortages could disturb pathways like the Sonic Hedgehog protein system, which regulates brain growth.

These findings point to a potential therapeutic target: reducing p-cresol and HPHPA to restore healthy CoA and dopamine balance.

Some physicians are already testing gut-focused therapies. One protocol involves a short course of metronidazole, followed by two months of rifaximin, an antibiotic that acts only inside the gut. Clinical results are often encouraging, with noticeable improvements in autistic symptoms.

But the journey is not always smooth. At the start of treatment, children often experience a temporary worsening of symptoms—a reaction similar to the Herxheimer effect, known from infectious diseases. As harmful microbes die off, they release toxins that briefly intensify inflammation before the body stabilizes and recovery begins.

This temporary setback may actually be a sign that the gut-brain system is recalibrating. For families, it highlights both the complexity and the promise of microbiome-based approaches in autism.

The Fungal Hypothesis in Autism

Among the many controversial topics in autism research, one of the most debated is the possible connection between fungal infections and autism spectrum disorders (ASD). Reports of children showing improvements—or in rare cases, a full recovery—after antifungal treatments have fueled both curiosity and skepticism. Medications such as fluconazole, nystatin, and amphotericin B have been prescribed "off-label" in this context, even though solid scientific evidence is still limited.

The idea dates back to the 1980s, when researchers noticed that repeated antibiotic use, especially in children treated for ear infections, disrupted gut bacteria and allowed yeast like Candida albicans to grow excessively. This overgrowth produced toxins that entered the bloodstream, altered immunity, and potentially affected the brain, leading to behaviors seen in autism.

One proposed mechanism is that Candida cells release ammonia. In the presence of propionic acid—a short-chain fatty acid already linked to autism—ammonia may transform into beta-alanine. Structurally, beta-alanine resembles the neurotransmitter GABA but with a slight difference. It can cross the blood-brain barrier and interfere with GABA signaling, possibly disrupting the brain's balance of excitation and inhibition. Abnormal GABA function has long been suspected in autism. Candida may also consume tryptophan, a building block of serotonin, potentially altering neurotransmitter balance.

Not all autistic children have Candida overgrowth, but some research highlights intriguing links. A 2000 open-label study treated 23 autistic children with nystatin. Tests showed high levels of compounds tied to Aspergillus species. After just 10 days on antifungal therapy, these compounds dropped significantly, suggesting fungal presence in the gut. Other researchers studied environmental conditions that favor mold growth. For example, a Harvard team in 2008 linked prenatal exposure to hurricanes and floods in Louisiana to higher autism prevalence, possibly due to mold contamination in homes.

Further evidence came in 2009 from California, where children exposed to mold toxins at home showed more severe neurological symptoms than unexposed autistic children. A Bulgarian study in 2019 added to the debate, showing unusual forms of filamentous fungi in the blood of nearly all autistic children studied, as well as in their mothers, but absent in controls. Antibodies against Aspergillus were also elevated, suggesting chronic fungal activity.

Perhaps the most striking account appeared in 2020. A boy with autistic symptoms and markers of Aspergillus in his gut was first treated with Saccharomyces boulardii, a probiotic yeast. The child developed a Herxheimer reaction, a temporary worsening that can occur when microbes die off and release toxins. Later, with careful monitoring, he was given itraconazole. Over three months, his autistic symptoms disappeared, and he made remarkable progress in academics, sports, and music. While just a single case, it sparked renewed debate.

Critics argue that such reports are anecdotal, and rigorous trials are lacking. Still, the hypothesis remains attractive because it suggests possible interventions. Environmental exposure is another angle. Wet housing, damaged by storms or poor ventilation, can host hidden colonies of Aspergillus or other molds. Parents have reported discovering mold in their children's bedrooms, later questioning its role in autism onset.

From a medical perspective, diagnosing fungal infections in autistic children is challenging. Standard tests are limited, especially in the sinuses or gut. Clinicians sometimes rely on therapeutic trials: if symptoms improve—or worsen—after antifungal treatment, it may suggest fungal involvement. However, antifungal drugs carry risks, and treatment must always be supervised.

Why does this matter? Because autism has often been viewed as either genetic or environmental, but the fungal hypothesis brings these together. Antibiotics, diet, stress, and living conditions can reshape the microbiome, tilting the balance toward fungal overgrowth. These changes might then interact with genetic vulnerabilities, immune responses, and even epigenetic regulation, amplifying autistic symptoms.

The controversy reflects a bigger truth: autism is not caused by a single factor. Instead, it emerges from complex interactions between genes, microbes, toxins, and the immune system. Even if fungi are just one piece of the puzzle, exploring their role could open new diagnostic and therapeutic strategies.

Looking forward, researchers emphasize the need for advanced tools such as next-generation sequencing, metabolomics, and computational modeling to identify fungal signatures and their interactions with the host. Controlled clinical trials could then test whether antifungal therapies—used carefully and selectively—can provide lasting benefits for subgroups of autistic children.

The story of fungal involvement in autism may still be incomplete, but it highlights how open-minded exploration, often led by persistent parents and courageous researchers, can challenge orthodox views. Whether or not fungi prove to be a central cause, their potential influence on the gut-brain axis is too important to ignore. For families, the message is clear: the environment matters, the microbiome matters, and sometimes, hidden factors like fungi may hold unexpected clues to understanding autism.

The Brain-Gut Connection and the Vagus Nerve in Autism

When we talk about autism, we often focus on the brain. But over the past two decades, science has revealed something surprising: the gut and its connection to the brain may play a much bigger role than anyone imagined. At the center of this story lies one of the most fascinating nerves in the human body – the vagus nerve.

The autonomic nervous system is responsible for the body's automatic functions, like breathing, digestion, and heart rate. It has three parts: the sympathetic system (the "fight or flight" system), the parasympathetic system (the "rest and digest" system), and the enteric nervous system (sometimes called the "second brain" of the gut). The vagus nerve is the main highway of the parasympathetic system, stretching from the brain down through the chest and abdomen, linking the brain to the heart, lungs, and digestive tract.

What makes the vagus nerve so special is that 80 percent of its fibers carry information from the organs to the brain, not the other way around. In other words, the gut is constantly talking to the brain. Scientists now believe that this constant "conversation" may help explain some of the social and sensory challenges found in autism.

The gut itself is lined with a network of more than 500 million neurons – more than the spinal cord. This "enteric nervous system" makes and releases many of the same neurotransmitters found in the brain, such as serotonin and dopamine. In fact, about 90 percent of serotonin and half of dopamine are made in the gut. No wonder researchers call it the body's "second brain."

The vagus nerve connects this second brain to the first. But the signals it carries are subtle. Nerve endings in the gut don't touch the microbes directly; instead, they detect chemicals released by bacteria or by gut cells reacting to food. For example, when food enters the intestine, special endocrine cells release serotonin, which activates vagus nerve fibers. Other gut hormones, like cholecystokinin (CCK) or peptide YY, also signal the brain this way.

Bacterial byproducts can also reach the vagus nerve. Short-chain fatty acids, such as butyrate, are made when gut bacteria digest fiber. These molecules can stimulate vagus nerve endings and, in turn, influence mood, stress, and even inflammation. Stress itself has the opposite effect: it weakens vagus activity and disrupts the gut, creating a feedback loop that may worsen symptoms in autistic children.

In the mid-1990s, Dr. Stephen Porges proposed a new way of looking at the vagus nerve, called the Polyvagal Theory. He argued that the vagus has two branches: a "ventral" branch that supports calmness, social interaction, and feelings of safety, and a more primitive "dorsal" branch that triggers shutdown or immobility when danger feels overwhelming. According to Porges, autistic individuals often struggle to access the ventral branch. Social situations, which require calm and open engagement, can instead feel threatening, leading to stress, sensory overload, or withdrawal.

Research supports parts of this view. Studies of heart rate variability – a measure of vagus nerve function – show that children with autism have slower or weaker vagal responses compared to neurotypical peers. This difference often becomes clear around 18 months of age, when autism symptoms also begin to appear. Other studies have shown reduced levels of certain brain receptors that respond to acetylcholine, the neurotransmitter released by the vagus nerve. This suggests the vagus is not only involved but may be an important target for therapy.

In 2019, a striking study from Baylor College of Medicine revealed just how direct the gut-vagus-brain link could be. Scientists gave autistic mouse models a probiotic called Lactobacillus reuteri. Social behavior improved, but when researchers cut the vagus nerve, the effect disappeared. It turned out that the probiotic was stimulating the vagus to send signals to the brain's hypothalamus, which then boosted oxytocin – the so-called "love hormone" that promotes bonding and trust. For mice lacking vagus signaling, the probiotic no longer worked. The lesson was clear: the vagus nerve acts as a key messenger between the gut microbiome and the social brain.

The therapeutic possibilities are exciting. Non-invasive ways of stimulating the vagus nerve are already being tested. Slow, deep breathing exercises strengthen vagal tone. Gentle massage around the ear – where branches of the vagus nerve are close to the surface – can activate it. Even singing, humming, or gargling create vibrations that stimulate vagus pathways. Some researchers are exploring transcutaneous electrical stimulation, where mild currents applied to the ear improve vagus function. Early results suggest benefits for anxiety, depression, epilepsy, and, potentially, autism.

Polyvagal Theory also reshapes how we think about autism therapies. If the vagus nerve helps regulate feelings of safety and connection, then approaches that create calm, predictable, and supportive environments may literally help the nervous system shift into a state more open to social engagement. This could explain why therapies that combine routine, sensory regulation, and safe relationships often show benefits.

Still, the science is young. Not all experts agree with Polyvagal Theory, and more clinical trials are needed to confirm whether vagus nerve stimulation can consistently improve autism symptoms. Yet the pieces are coming together: the gut microbiome, the vagus nerve, and the brain form a three-way partnership that seems central to human health.

The big picture is that autism cannot be reduced to just "genes" or "behavior." The body's communication networks – the gut, the immune system, the brain, and the vagus nerve that links them – must be seen as part of the story. This doesn't mean there is a simple fix. But it does suggest that treatments focusing on restoring balance, whether through diet, probiotics, stress reduction, or gentle vagus stimulation, may offer real hope.

For parents, this line of research can be both challenging and inspiring. On one hand, it highlights how complex autism really is. On the other hand, it shows that the biology of autism is not frozen or predetermined. The gut and the brain are in constant dialogue. The vagus nerve carries that conversation every second of every day. And if we can learn to guide or strengthen that conversation, we may be able to open new doors for therapy, resilience, and human connection.

Autism and the Thyroid Connection

Normal thyroid function is essential for healthy brain development and cognitive abilities. Over the past decades, researchers have found more links between thyroid disorders and various psychiatric or neurological conditions. These include certain cases of intellectual disability, psychosis, panic disorders, anxiety, depression, and even schizophrenia. The exact nature of this relationship is still unclear, but the evidence is growing stronger.

When it comes to autism, many scientists have asked whether thyroid problems might play a role. What makes this theory intriguing is that many symptoms seen in children with autism overlap with those of thyroid dysfunction. Constipation, digestive troubles, sleep disturbances, delayed language, hyperactivity or unusual fatigue, bloated stomachs, prolonged jaundice, eating issues, brittle hair and nails, cold hands and feet, anxiety, extreme fears, trouble focusing, and weak muscles—all of these can appear in both conditions. Some experts even suggest that certain cases of autism may actually be hidden hypothyroidism.

One of the most studied risk factors is maternal hypothyroidism, which means low thyroid function during pregnancy. Research has shown that when thyroid-stimulating hormone (TSH) levels are high in pregnant women, the risk of having a child with autism increases. For children already diagnosed with autism, other risk markers have been identified, such as high levels of anti-thyroid peroxidase antibodies (anti-TPO) or a family history of autoimmune thyroid disease.

The reason thyroid health is so important during pregnancy is that maternal thyroid hormones guide the development of the nervous system in the fetus. Hormones like triiodothyronine

(T3) and thyroxine (T4) regulate genes critical for brain growth, including BDNF and reelin, both of which have been associated with autism. Some studies show that abnormal thyroid hormone levels at birth are linked to a higher chance of developing autism spectrum disorder.

Another factor to consider is the presence of folate receptor alpha autoantibodies (FRAA), which are found in high levels in many autistic individuals. These antibodies can interfere with TSH, raising its levels and making the thyroid gland less responsive. They may also affect the sensitivity of the hypothalamic-pituitary-thyroid axis, adding another layer of complexity.

Environmental toxins also enter the picture. Substances like polychlorinated biphenyls (PCBs), lead, mercury, and dioxins can disrupt both the brain and the thyroid. This dual impact could help explain why exposure to such chemicals during pregnancy is sometimes linked to higher autism risk.

In clinical practice, thyroid evaluation often begins and ends with a TSH test. But this approach may miss subtle dysfunctions. That is why some doctors recommend a more complete profile, measuring free T3, free T4, reverse T3, and checking for thyroid autoimmunity with anti-TPO and anti-thyroglobulin antibodies. This detailed testing can reveal hidden thyroid imbalances that may show up only as behavioral or cognitive symptoms, easily mistaken for autism traits.

By identifying these hidden dysfunctions early, doctors can propose personalized treatment plans that may reduce suffering and avoid misdiagnosis. While not all autism cases are tied to thyroid problems, the overlap is compelling enough to warrant careful screening. As research continues, the thyroid may prove to be one of the missing puzzle pieces in understanding the complex biology of autism.

Carnitine and Autism: A Hidden Connection

Carnitine is a small molecule found in almost every cell of the human body. It is especially concentrated in the muscles, where it plays a key role in turning fat into energy. Carnitine comes from two sources: what the body produces and what we get from food. Our body makes it from amino acids like lysine and methionine, with the help of vitamins and minerals such as vitamin D, iron, vitamin B6, and niacin. Even though we only produce small amounts each day, carnitine's role is essential for many functions.

Scientists have discovered that carnitine is more than just a helper for energy. It is also a protector of the brain and body. It works as an antioxidant, helping the body fight harmful free radicals and reducing inflammation. Carnitine supports the mitochondria, the "power plants" of our cells, which are often weakened in children with autism. By supporting mitochondrial energy production, carnitine helps the brain and muscles function more efficiently.

Carnitine also plays a role in the brain itself. It can increase dopamine and GABA release—two key brain chemicals involved in mood, behavior, and focus. It also helps produce acetylcholine, a neurotransmitter important for memory and learning. In addition, carnitine influences gene expression, glucose metabolism, and even detoxifies certain drugs and chemicals. Its many roles explain why researchers have begun to link it with autism.

The link between autism and carnitine became clearer through three main observations. First, many children with autism show signs of mitochondrial dysfunction, which carnitine helps regulate. Second, studies found that children with autism often have lower levels of carnitine than their peers. Third, some genetic mutations that affect carnitine metabolism are more common in autistic children. Together, these findings suggest that carnitine deficiency may be a hidden factor in autism.

One large study reported that more than 80 percent of autistic children tested had low levels of total and free carnitine. Another study led by Dr. Richard Frye identified unusual patterns of acyl-carnitines–forms of carnitine that carry fatty acids into mitochondria–similar to patterns seen in animal models of autism. These abnormalities suggest that the body's ability to use fat for brain energy may be impaired.

Because of these results, carnitine supplementation has attracted attention. Doses up to 200 mg per kilogram of body weight per day have been tested, and the treatment was safe and well tolerated. Several studies reported improvements in autism scores, with better language, sociability, and reduced repetitive behaviors. For some children, the changes were dramatic.

Genetics may also explain the connection. Researchers discovered that a mutation in the TMLHE gene, which controls the first step of carnitine production, is present in some autistic children. Supplementing carnitine in these cases led to major improvements in communication and social interaction. Another gene, SLC7A5, which also influences carnitine transport, may explain why autistic children respond differently to treatments like risperidone, a common medication.

From a clinical perspective, measuring carnitine levels in autistic children may reveal hidden deficiencies. Correcting them can change developmental outcomes. In my own medical practice, I have seen children show significant gains after targeted carnitine therapy. In some cases, children even lost their autism diagnosis entirely after supplementation.

Among the different forms of carnitine, acetyl-L-carnitine (ALC) seems especially promising for the brain. Unlike regular carnitine, ALC easily crosses the blood-brain barrier. Once inside, it boosts mitochondrial energy, reduces oxidative stress, and increases acetylcholine production. These effects can improve learning, memory, emotional regulation, and social skills. Clinical studies and real-world observations confirm that ALC improves attention, reduces repetitive behaviors, and enhances overall neurological outcomes.

The story of carnitine and autism is not just about one molecule. It is a reminder that autism may have biological roots that are often overlooked. While genetics and the environment both play roles, metabolism—how the body turns food and molecules into energy—may be equally important. By addressing metabolic weaknesses, such as low carnitine, doctors and researchers may unlock new ways to support autistic children.

In the future, carnitine supplementation could become part of a broader, personalized treatment approach for autism. Combined with nutritional strategies, microbiome care, and other targeted therapies, it may help improve quality of life and developmental progress. While more large-scale studies are needed, the evidence so far suggests that carnitine is far from a minor player. For some children, it may be the missing piece of the puzzle.

Paracetamol and Autism: A Controversial Link

In a surprising and dramatic statement, U.S. President Donald Trump recently raised alarms about the potential risks of paracetamol (acetaminophen), one of the world's most widely used painkillers. His words sparked heated debates across the medical community, drawing attention to a controversial and growing body of research suggesting that paracetamol might be linked to autism spectrum disorders (ASD). This discussion has reignited concerns about the safety of a drug long considered harmless at therapeutic doses.
For decades, paracetamol has been one of the most common over-the-counter drugs, praised for its ability to relieve pain and reduce fever. Doctors and pharmacists alike presented it as a safe option for children, pregnant women, and anyone needing quick relief. But today, that image of safety is under fire. A growing number of scientists and physicians are questioning whether this drug, so deeply embedded in daily life, could play a role in serious conditions like autism spectrum disorder (ASD).

The controversy began with unsettling research. Paracetamol, first derived from an industrial chemical called aniline, replaced older drugs like acetanilide and phenacetin, which were abandoned due to dangerous side effects. While paracetamol has avoided the same fate, its exact mechanism of action is still not fully understood. Scientists now know that paracetamol indirectly activates cannabinoid receptors in the brain. Its metabolite, AM404, prevents the reuptake of anandamide, a natural brain chemical, which influences both pain and mood. Blocking these receptors completely stops paracetamol from working. This discovery has led some researchers to wonder: what else is this drug doing to the developing brain?

One of the first to raise the alarm was Dr. Stephen T. Schultz from the University of California, San Diego. In 2008, Schultz and his team published a case-control study suggesting that children who took paracetamol at 12 and 18 months of age were up to eight times more likely to develop autism compared to those who did not. Even more striking, when looking at autistic children with developmental regression, the risk jumped to 21 times. This was a startling finding that could not be ignored.

In a later paper in 2010, Schultz made another provocative observation: after the medical community shifted from prescribing aspirin to paracetamol in the early 1980s, autism diagnoses began to rise dramatically. Aspirin had fallen out of favor because of concerns about Reye's syndrome, a rare but severe condition. Pharmaceutical companies encouraged the shift, presenting paracetamol as safer for children. According to Schultz's analysis, autism rates only dipped in 1982 and 1986, years when paracetamol sales in the United States dropped sharply due to high-profile scares about cyanide-laced capsules. The coincidence raised troubling questions.

Together with Dr. Georgianna G. Gould of the University of Texas, Schultz proposed a detailed scenario of how paracetamol might influence autism risk. Their theory suggested that when a child with fever takes paracetamol, the drug boosts activity in the brain's endocannabinoid system, bringing temporary relief. But this same system also helps regulate immunity and neural development. Overstimulation could weaken the immune response, prolong illness, and disrupt brain development. Later, the system may overcorrect by shutting down too much, leaving developing neurons without the signals they need. The result: altered brain development and behavior consistent with autism spectrum disorder.

Supporting this idea is a curious observation made by many parents and doctors: some autistic children become temporarily more sociable and communicative during a fever. Scientists suggest this happens because fever triggers a surge of anandamide, briefly restoring balance in the endocannabinoid system. The momentary improvement in social behavior fits with the theory that paracetamol disrupts this delicate system.

Beyond the brain, paracetamol also poses challenges for the liver. In young children, the drug is usually processed through sulfation. But if this pathway is weak, the drug instead produces a toxic byproduct called NAPQI, which can only be neutralized by glutathione, a key antioxidant. Excess paracetamol use can quickly drain both sulfate and glutathione stores, weakening the body's ability to detoxify other harmful substances. Many researchers now believe that children with autism often have impaired detoxification systems, making them especially vulnerable.

Parents often report that autistic behaviors worsen after their children eat certain foods—wheat, corn, chocolate, bananas, and dairy among them. These foods are rich in phenolic compounds that also require sulfation for removal. If the liver cannot handle both dietary phenols and paracetamol, neurotransmitters like dopamine and norepinephrine may accumulate in the brain, producing toxic effects. Several studies have shown lower levels of the enzyme PST (phenolsulfotransferase) in autistic children, supporting the idea of a weak detox system.

The evidence is not limited to children. Several studies have raised alarms about paracetamol use during pregnancy. In 2019, a major prospective cohort study published in JAMA Psychiatry by Dr. Xiaobin Wang and colleagues at Johns

Hopkins analyzed 996 mother-child pairs. They found a dose-dependent relationship between umbilical cord levels of paracetamol metabolites and later diagnoses of ADHD and autism. By age nine, nearly 26% of the children had ADHD, 6.6% had autism, and 4.2% had both. Children with the highest exposure levels were more than three times as likely to be diagnosed with autism.

Other epidemiological studies have echoed these findings, linking prenatal exposure to both autism and ADHD. While none can prove causation, animal experiments show that immature brains exposed to paracetamol suffer long-term changes. The exact biological pathways remain unclear, but the consistency of the associations worries researchers.

The story of paracetamol does not end with autism. Scientists have also found potential links to asthma and male reproductive issues like cryptorchidism. These associations suggest that the drug's long-term effects may extend far beyond pain relief and fever reduction. Once celebrated as a nearly risk-free medicine, paracetamol may carry hidden costs that society is only beginning to understand.

Interestingly, paracetamol's acceptance was not immediate. When first introduced in France in the 1950s, doctors were skeptical. Many doubted its effectiveness and questioned its safety, especially since it lacked the anti-inflammatory benefits of aspirin. Only after studies showed it was safer for patients with stomach ulcers did paracetamol gain widespread acceptance. Today, it dominates pharmacy shelves worldwide. Yet in light of emerging evidence, experts are calling for a careful reexamination of its risks, particularly for pregnant women and young children.

The growing body of research does not mean parents should panic, nor does it prove that every case of autism is linked to paracetamol. But the pattern of evidence—from epidemiological studies, biochemical insights, and clinical observations—is too strong to dismiss. The myth of paracetamol as a completely safe drug is crumbling. What began as a wonder medicine of the 20th century may turn out to be a hidden risk factor for one of the most pressing neurodevelopmental challenges of our time.

Autism and Calcium: A Hidden Connection

Calcium is the most abundant mineral in the human body. Everyone knows it as the building block of strong bones and teeth, but its role extends far beyond that. Calcium is essential for growth in children, maintaining bone strength in adults, and preventing osteoporosis in older people. It also supports muscle contraction, nerve communication, and blood clotting–functions without which life would not be possible.

Most of the body's calcium is stored in the bones in the form of calcium phosphate. Only a small portion, about 1%, circulates in soluble form as Ca2+ ions. Cells maintain calcium inside at an extremely low concentration (about 10^{-7} mol/L), compared to the much higher level outside (about 10^{-3} mol/L). This difference is vital, because too much calcium inside a cell can be deadly.

Cells must be able to quickly change their internal calcium levels in response to signals from outside. This dynamic response relies on two main factors: the entry of calcium from the extracellular environment and its release from an internal storage system, the endoplasmic reticulum. This structure acts as a calcium reservoir, releasing ions when needed. To maintain balance, the cell uses a network of pumps, channels, and binding proteins. Among them, the IP3 receptors stand out, because they regulate the release of calcium from the endoplasmic reticulum.

But how does calcium connect to autism? More and more research suggests that calcium signaling may play a role in autism spectrum disorders (ASD). One of the most striking examples comes from Timothy syndrome, a rare genetic disease caused by severe mutations in the CACNA1C gene. This gene controls calcium channels called Cav1.2, found in both the brain and the heart. People with this mutation often present with heart problems and symptoms of autism. Other calcium channel disorders, known as "channelopathies," have also been linked to autism.

In 2015, Dr. J. Jay Gargus and his team at the University of California, Irvine, published groundbreaking research on this topic. By studying skin cell biopsies from patients with three different genetic conditions associated with autism—the Fragile X syndrome and tuberous sclerosis complex (mutations TSC1 and TSC2)—they found that calcium signaling was deeply altered. In all three conditions, the IP3 receptors that control calcium release inside the cells were less active than normal. This defect, identified within the endoplasmic reticulum, could explain not only neurological issues, but also immune and digestive problems often observed in autism.

On another front, Italian professor Antonio Persico suggested that the autistic brain may have both too much calcium and excessive calcium signaling. This idea points to a dual problem: sometimes calcium activity is too low, and other times it is too high. Either way, the balance is broken. Such dysregulation could be used as a diagnostic marker and, more importantly, as a therapeutic target.

The picture becomes even clearer when we look at glutamate, the brain's main excitatory neurotransmitter. In autism, the NMDA receptors—channels that open in response to glutamate—may become overactive. This allows too much calcium to flow into neurons. When that happens, crucial processes such as synaptic plasticity, cell signaling, and network regulation are disturbed. The result can be neuronal damage, oxidative stress, and chronic brain inflammation.

These disruptions interfere with brain development, leading to the behavioral and cognitive challenges typical of autism.

The idea of targeting calcium signaling as a therapy is now gaining attention. By carefully modulating NMDA receptors and calcium channels, researchers hope to restore balance in the autistic brain. This approach could protect neurons, reduce inflammation, and improve learning and social abilities.

Calcium, once seen only as the mineral for bones, is proving to be a key player in the brain. Its mismanagement at the cellular level may be one of the hidden threads connecting autism's many puzzling symptoms. As science advances, calcium signaling could become both a tool for early detection and a path toward new treatments for autism spectrum disorders.

Autism and Microglia

One of the most intriguing pieces in the autism puzzle lies in the role of microglia. Glial cells, which surround and support neurons, make up more than half of all brain cells and almost half of the brain's volume. In the central nervous system, there are two main groups: macroglial cells (astrocytes, oligodendrocytes, and ependymal cells) and microglial cells.

Microglia belong to the macrophage family and act as the brain's main active immune defense, representing around 12 to 15 percent of all brain cells. They form a vast three-dimensional network, with each cell monitoring its own territory. Their small bodies extend thin, mobile branches that scan their environment like patrols. Through this constant surveillance, microglia detect external signals, integrate them, and respond in ways that help maintain brain balance and health.

It is becoming increasingly clear that the microglial response is highly variable and critical for brain development, worsening disease, or even healing. One of their key roles is synaptic pruning – the process of eliminating unused synapses so that the brain can operate more efficiently. This pruning is essential for learning, especially from early childhood through adulthood. Too little or abnormal pruning has been linked to developmental disorders, including autism.

Glial cells are also central to brain plasticity, the idea that the brain is not a static organ but one that constantly reshapes itself. Contrary to older beliefs, the brain can create, dismantle, and reorganize neurons and synapses throughout life.

Synaptic pruning is tightly connected to this plasticity, ensuring that unused synapses fade while active ones grow stronger. This reshaping means that the brain's map is always changing, largely thanks to microglia.

When confronted with threats, microglia become activated and change shape. They retract their branches, adopt a more mobile form, and express specific immune markers. This is known as the pro-inflammatory or M1 state. In this state, microglia release pro-inflammatory molecules such as interferons, IL-1, IL-6, IL-12, and TNF-alpha.

However, microglia can also adopt an anti-inflammatory or M2 state. In this mode, they release protective molecules like IL-10 and TGF-beta that help repair tissues. Some researchers now suspect that in autism, microglia remain stuck too often in the M1 state. This could block synaptic pruning and contribute to learning and cognitive problems seen in autism.

Many believe that microglia only become inflamed if invaders directly enter the brain. In reality, they act more like an 'immunostat,' a term coined by Dr. Rodney W. Johnson at the University of Illinois. They can react to signals from distant parts of the body, making them sensitive to much more than direct brain infections.

Scientists have described several ways inflammation in the body can activate microglia. Sometimes, inflammatory molecules cross the blood-brain barrier directly. Other times, they attach to receptors on blood vessel cells, which then signal nearby microglia to release their own inflammatory molecules. Gut inflammation may also reach the brain via the vagus nerve, which can carry immune signals that influence behavior and cognition. Even slower pathways exist, with inflammatory molecules diffusing through brain fluids and activating microglia far from the original site.

Abnormal microglial activation and faulty synaptic pruning have been documented in autism through both animal studies and human brain tissue. These issues may start during pregnancy or in early childhood. Animal experiments show that maternal immune activation – such as an infection during pregnancy – can increase inflammatory molecules like IL-6, leading to changes in the newborn's brain that resemble autism.

Cytokines like IL-1, IL-6, and TNF-alpha can cross into the fetal brain and directly alter its development. In animal studies, exposing pregnant mothers to compounds mimicking viral or bacterial infections led to autistic-like behaviors in offspring. Microglial activation appears to be one of the key mechanisms behind this effect.

In 2020, researchers from the Scripps Research Institute published a study in *Nature Communications* that added another piece to the puzzle. They found that when microglia produced too much of certain proteins, male mice showed autistic-like behaviors. This was not seen in female mice or in cases where other brain cells overproduced proteins. One protein, EIF4E, which boosts the production of other proteins, was particularly important. Overproduction made microglia swell and lose mobility, preventing them from pruning synapses properly.

One of the challenges in autism research is that human microglia differ from those in animals. This may explain why animal studies don't always mirror human symptoms. New technologies, like human brain organoids, now allow scientists to study microglia in lab-grown mini-brains, helping close this gap.

Given their growing importance, many researchers now see microglia as promising therapeutic targets for autism. Treatments that calm overactive microglia or restore their balance could improve symptoms. For instance, anti-inflammatory drugs that act directly on microglia or molecules that adjust cytokine signaling could help rebalance brain immunity. By restoring normal pruning, such treatments could improve learning and social behaviors.

Some substances are already being studied for their ability to regulate microglia. Luteolin, a natural antioxidant, can block microglial activation and reduce toxic inflammation. Minocycline, an old antibiotic, also has anti-inflammatory and neuroprotective effects. Suramin, once used to treat African sleeping sickness, blocks a signaling pathway involved in microglial activation. Vitamin D, which acts like a neurosteroid, reduces inflammatory signals released by microglia. Palmitoylethanolamide, a natural fatty acid amide, may also reduce inflammation by acting on cannabinoid receptors.

Other strategies include compounds like EGCG (from green tea), probiotics, or targeting the kynurenine pathway – all aiming to restore healthy brain chemistry. By reducing excessive inflammation and promoting balanced brain-immune interactions, these approaches could improve synaptic function, learning, and flexibility. Over time, they may even improve social and cognitive abilities in people with autism.

Although research is still young, evidence strongly suggests microglia play a critical role in autism. Understanding how their dysfunction contributes to the condition could unlock new treatments, offering hope for better outcomes in the future.

Vitamin D: The Missing Piece of the Autism Puzzle

Over the past two decades, Vitamin D has moved from being seen simply as a nutrient for bone health to being recognized as a powerful hormone-like substance with far-reaching effects. Researchers now call it a "neurosteroid" because of the way it influences the brain. And among the many conditions linked to Vitamin D, autism spectrum disorders (ASD) stand out as one of the most debated and intriguing.

If there is one environmental factor that could explain much of the autism epidemic, it may well be Vitamin D deficiency.
The idea may sound bold, but growing evidence from epidemiology, genetics, immunology, and neuroscience has been converging on the same point: Vitamin D may be a key regulator of brain development, immune balance, and even social behavior.

The Hill Criteria and Autism

The famous British epidemiologist Bradford Hill proposed nine criteria to judge whether a correlation suggests causation. They include strength of association, consistency across studies, dose-response relationships, biological plausibility, and more. When these criteria are applied to Vitamin D and autism, many researchers argue that they fit.
Studies across different populations show the same link. Animal experiments confirm biological plausibility. Clinical trials reveal benefits of supplementation. And the timing makes sense: a deficiency in pregnancy or early life precedes the onset of symptoms. All of this suggests that Vitamin D deficiency could be more than coincidence–it could be causal.

How Vitamin D Works in the Brain

Vitamin D activates genes that control enzymes such as tryptophan hydroxylase 2 (TPH2). This enzyme helps convert tryptophan, an amino acid from food, into serotonin in the brain. Serotonin is a neurotransmitter critical for wiring the developing brain and regulating social interaction. Without enough Vitamin D, the brain may fail to produce enough serotonin, leading to problems in behavior and cognition.

At the same time, Vitamin D inhibits the gene for tryptophan hydroxylase 1 (TPH1), which produces serotonin in the gut.
Too much serotonin outside the brain is linked to inflammation. This mechanism explains a paradox that has puzzled autism researchers for decades: children with autism often show low serotonin in the brain but high serotonin in the blood. Vitamin D provides a unifying explanation.

Boys, Girls, and Hormones

Autism affects boys more than girls, a fact that has long puzzled scientists. Estrogen appears to protect the female brain, boosting serotonin levels. Vitamin D raises estrogen in the placenta and brain, suggesting that boys may be especially vulnerable when Vitamin D is lacking. This helps explain the gender imbalance in autism prevalence.

Immune System and Inflammation

Mothers of children with autism are more likely to carry antibodies that target the fetal brain. Vitamin D is a master regulator of the immune system. It promotes regulatory T-cells, reduces inflammatory cytokines like IL-6 and TNF-alpha, and helps maintain tolerance to the developing fetus. A lack of Vitamin D could therefore set the stage for autoimmunity during pregnancy, with lifelong consequences for the child.

Heavy Metals, Pesticides, and Pollution

Debates about autism often focus on environmental toxins such as mercury, lead, or pesticides. These substances indeed can harm the brain, but their impact may be magnified when Vitamin D is low. That's because Vitamin D regulates glutathione, the body's main detoxification molecule. Without enough Vitamin D, glutathione levels drop, making it harder to clear toxins. Air pollution also reduces Vitamin D by blocking UVB sunlight, further compounding the risk.

The Gut, the Microbiome, and Vitamin D

The gut-brain axis has become one of the hottest areas of autism research. Vitamin D plays a major role here too.

It strengthens the gut barrier, influences which microbes thrive, and shifts the immune system toward an anti-inflammatory state. Deficiency can disrupt all of these processes, creating a "perfect storm" for neurodevelopmental disorders.

Pregnancy, Birth, and Risk Factors

Low Vitamin D in mothers is linked to complications like preeclampsia, gestational diabetes, premature birth, and low birth weight. Each of these conditions is also associated with higher autism risk. Similarly, cesarean deliveries are more common when mothers are deficient in Vitamin D. The parallels are striking.

Early Clues from Rickets and Williams Syndrome

Historical reports from the early 20th century describe children with rickets, a disease caused by Vitamin D deficiency, who displayed behaviors we would now call autistic. When they were treated with high-dose Vitamin D, their symptoms improved. On the other end of the spectrum, children with Williams syndrome, a rare genetic disorder marked by excess Vitamin D, show hypersociability—the opposite of autism. These extremes suggest Vitamin D levels may powerfully influence social behavior.

Modern Clinical Trials

Several clinical trials have tested Vitamin D supplementation in children with autism. Results show improvements in language, social interaction, and reduction of repetitive behaviors. In Egypt, a study on 122 children found that 75% showed significant improvement after supplementation, with no toxicity. Another randomized trial confirmed benefits with doses up to 5,000 IU per day. A Chinese meta-analysis of 452 publications concluded that Vitamin D improves scores on gold-standard autism rating scales.

Eight Ways Vitamin D May Protect the Brain

Researchers have outlined at least eight mechanisms by which Vitamin D supports brain health in autism:

1. DNA repair and protection against cellular stress.
2. Anti-inflammatory effects reducing chronic brain inflammation.
3. Modulation of autoimmunity and reduction of harmful antibodies.
4. Boosting regulatory T-cells to calm the immune system.
5. Increasing neurotrophins that support neuron growth and plasticity.
6. Protecting mitochondria, the cell's energy factories.
7. Enhancing glutathione production for detoxification.
8. Supporting neurotransmitters and synaptic plasticity.

Practical Implications

For clinicians, parents, and policymakers, the implications are clear. Maintaining optimal Vitamin D levels—between 40 and 60 ng/ml—may reduce autism risk and improve outcomes. Supplementation is inexpensive, safe when monitored, and potentially life-changing. Leading experts like Dr. John

Cannell argue for systematic supplementation of pregnant and breastfeeding women, as well as young children.

Rethinking Old Fears

Vitamin D has long been feared for toxicity risks, but those fears stem from outdated studies. Modern research shows that doses up to 10,000 IU per day are safe for most people when monitored. The real danger, experts argue, lies in deficiency—not excess. Vitamin D is not just a vitamin; it is a hormone that regulates thousands of genes.

Conclusion

As autism rates continue to rise, the search for answers grows more urgent. While genetics play a role, no epidemic in history has been purely genetic. Environmental factors must be involved. And among them, Vitamin D stands out as a prime candidate: it is simple, measurable, and correctable. Ensuring that mothers and children get enough Vitamin D may not solve autism entirely, but it could prevent countless cases and ease the burden for millions of families.

Vitamin D, once seen as only the "bone vitamin," may prove to be the missing piece of the autism puzzle.

The Role of GABA and Glutamate Systems in Autism

Autism research has long focused on how the brain's chemical messengers, or neurotransmitters, might explain the condition's wide range of symptoms. Among these messengers, two stand out: glutamate, the brain's main excitatory neurotransmitter, and GABA (gamma-aminobutyric acid), which is derived from glutamate and acts as its calming counterpart. Together, they form a delicate balance between excitation and inhibition. When this balance breaks down, the effects can be profound.

Glutamate operates through different receptors, including NMDA (N-methyl-D-aspartate), AMPA, and metabotropic glutamate receptors (mGluR). Many studies suggest that NMDA and AMPA receptors are deeply involved in the biology of autism. On the other hand, GABA receptors change with age. During early brain development, GABA acts as an excitatory force, shaping how neurons grow, migrate, and connect. Later, in adulthood, GABA takes on its well-known inhibitory role, calming excessive activity in the nervous system.

The problem arises when these systems malfunction. A disrupted balance between glutamate and GABA can create a tilt toward hyperexcitability or excessive inhibition. In animal models, scientists have shown that disturbing this balance alters how information is processed in the brain and weakens social behavior. In children with autism, blood studies often show abnormal levels of GABA and glutamate. Typically, there is an increase in GABA, a higher glutamate-to-glutamine ratio, and a decrease in glutamine levels. This imbalance echoes patterns also found in epilepsy and schizophrenia, suggesting a shared biological link.

The consequences of this imbalance are not just numbers in a lab test. High glutamate activity, for example, is associated with brain inflammation. When glutamate receptors, especially NMDA receptors, are overactivated, they allow an uncontrolled flood of calcium into neurons. This process, known as excitotoxicity, damages cells and fuels oxidative stress and inflammation. In autism, such mechanisms may underlie repetitive behaviors, irritability, hyperactivity, and even obsessive-compulsive tendencies.

Scientists have explored drug treatments to fix this imbalance, though results remain mixed. Some drugs target the GABA system, such as bumetanide, arbaclofen, acamprosate, and valproate. So far, the clinical outcomes have been inconsistent, and more research is needed. Interestingly, both boosting and blocking NMDA receptors have shown benefits. The drug D-cycloserine, which stimulates NMDA receptors, has reduced social withdrawal and repetitive behaviors. Meanwhile, memantine, which blocks NMDA receptors, has eased irritability, hyperactivity, and stereotypies. These seemingly opposite approaches highlight just how complex the glutamate-GABA relationship is in autism.

But chemicals are not the only culprits. Everyday factors may fuel the imbalance. Diets high in food additives, such as monosodium glutamate (MSG), or heavily processed foods can raise glutamate activity. Stress, infections, and chronic inflammation also tip the scale. Deficiencies in nutrients like vitamin D, magnesium, or omega-3 fatty acids further weaken the brain's defenses, while environmental toxins worsen the problem.

Amid these challenges, some interventions look promising. Taurine, an amino acid naturally found in the body, has shown potential in calming glutamate-induced excitotoxicity. By controlling calcium entry into neurons, taurine helps prevent the chain reaction that leads to inflammation and neuronal damage. Scientists believe such natural compounds could eventually complement drug therapies, offering safer and more accessible treatment options.

One of the greatest obstacles in autism research remains the condition's enormous diversity. No two children experience autism in exactly the same way. This heterogeneity complicates clinical trials, as what works for one child may fail for another. Still, the mounting evidence around the GABA and glutamate systems offers a compelling picture: autism may, at least in part, stem from a fragile balance between excitation and inhibition in the brain.

If this is true, then restoring that balance—through nutrition, lifestyle, medication, or a combination of these—could open new doors to understanding and treating autism. The story of glutamate and GABA is still unfolding, but it holds the promise of unlocking some of the deepest mysteries of the autistic brain.

Serotonin, Dopamine, and Autism

When scientists and doctors look into autism spectrum disorder (ASD), they often search for answers in the brain's chemistry. Two of the most important chemical messengers in this story are serotonin and dopamine. These neurotransmitters are not only crucial for communication between brain cells, but they also influence mood, learning, sleep, motivation, and social behavior. Understanding how serotonin and dopamine work, and how they sometimes malfunction, may be one of the keys to better understanding autism.

Serotonin: The Mood Regulator with a Broader Role

Serotonin, often called 5-HT, is one of the most famous neurotransmitters. Most people know it as the "happiness chemical" because it helps regulate mood and sleep. But its functions go far beyond that. Serotonin is also involved in shaping the developing brain. It influences how neurons grow, how they connect with each other, and how the brain forms its network of pathways. During development, serotonin plays a key role in processes like synapse formation, neuronal migration, and the wiring of brain circuits.

Interestingly, serotonin is not only in the brain. Large amounts of it are found in blood platelets. This connection between serotonin in the blood and serotonin in the brain has raised many questions about its link with autism.

High Serotonin in the Blood, Low Serotonin in the Brain?

One striking fact is that about 30 percent of people with autism have something called hyperserotonemia – elevated serotonin levels in the blood. At first, this might suggest that serotonin is too high in autism. But scientists now believe the picture is more complicated. The serotonin stored in blood platelets may not reflect what is happening in the brain. In fact, some research suggests that while blood serotonin is high, serotonin in the brain may actually be low. This paradox has puzzled researchers.

Animal studies support this idea. In certain mouse models of autism, serotonin in the brain is reduced even though overall serotonin activity in the body is high. This may explain why some children with autism have lower brain serotonin synthesis compared to neurotypical children.

The link between serotonin and autism is further supported by tryptophan depletion studies. Tryptophan is an amino acid the body uses to make serotonin. When tryptophan levels drop, brain serotonin also decreases, and in children with autism, symptoms often worsen. This suggests that low serotonin in the brain may indeed play a role in autism behaviors.

Genetic Clues in the Serotonin System

Genetics also point to serotonin's involvement. Variations in genes connected to serotonin signaling have been linked to autism. One gene in particular, the serotonin transporter gene (SERT or 5-HTT), has been widely studied. The transporter controls how serotonin is recycled in the brain. Certain polymorphisms (genetic variations) in SERT have been associated with autism. One variant, called Gly56Ala, appears more often in individuals with autism. This variant is linked to rigid, compulsive behaviors and sensory aversion.

Other genetic studies show reduced binding or expression of serotonin receptors, such as 5-HT1A and 5-HT2A, in key brain regions like the striatum and anterior cingulate cortex. Neuroimaging has also revealed lower serotonin transporter activity in brain areas important for emotion and decision-making. Together, these findings suggest that the serotonin system in autism is less responsive – either because of fewer receptors, weaker receptor binding, or altered transporter activity.

Serotonin, Aggression, and Empathy

Low serotonin has also been linked to aggression, which can be a challenging symptom in autism. Social provocation, frustration, or difficulties in empathy may trigger aggressive responses. Since serotonin helps regulate empathy and moral judgment, disruptions in this system could partly explain aggressive behavior in autism.

For example, studies show that drugs increasing brain serotonin, such as SSRIs (selective serotonin reuptake inhibitors), can make people more sensitive to the idea of harming others. Similarly, MDMA (ecstasy), which boosts serotonin release, is known to temporarily increase empathy. These effects suggest that serotonin shapes how people interpret social cues and regulate emotional responses. In autism, reduced serotonin activity might make it harder to read facial emotions, show empathy, or respond in appropriate way.

Dopamine: The Reward Messenger

While serotonin is often associated with mood and empathy, dopamine is better known for its role in motivation, movement, and reward. Dopamine helps us seek out pleasure, enjoy rewards, and stay motivated. It also plays a big role in how we learn from new experiences. Dopamine signals are particularly important in social interactions, where they help assign value to people, events, and relationships.

There are several major dopamine pathways in the brain. The nigrostriatal pathway controls movement. The mesocortical pathway links to thinking and decision-making. The mesolimbic pathway is central to reward and motivation. All of these circuits can be affected in autism.

Genetic Links to Dopamine and Autism

Several dopamine-related genes have been tied to autism. These include genes coding for dopamine receptors (D1, D2, D3), transporters, and enzymes involved in dopamine production. Many of these genes appear in the SFARI Gene database, a resource cataloging autism-associated genes. Their inclusion reflects strong statistical evidence of association, even though they also play roles in other disorders like ADHD. This overlap suggests dopamine is part of a broader web of brain chemistry problems.

Brain Imaging: A Window into Dopamine Function

Modern brain imaging has been especially helpful in showing how dopamine systems function differently in autism. Functional MRI (fMRI) studies show that the striatum of people with autism responds less to rewards, especially social ones. This could explain why social interactions, which are naturally rewarding for most people, may feel less motivating for individuals with autism.

Positron emission tomography (PET) studies also reveal altered dopamine activity. Some studies show increased dopamine transporter binding in certain brain regions, while others show reduced dopamine release in response to rewards. These changes disrupt the brain's ability to link social experiences with positive feelings, contributing to difficulties in social engagement.

Medication: What Works and What Doesn't

Currently and before the recent announcements about folinic acid (leucovorin), the only FDA-approved medications for autism target irritability and aggression, not the core symptoms. These drugs are risperidone and aripiprazole. Both act on dopamine receptors, either blocking them (as risperidone does) or partially stimulating them (as aripiprazole does). While they can reduce aggression and irritability in the short term, they come with significant side effects, including weight gain and movement problems. Importantly, they do not improve the main challenges of autism.

Lessons from Mice: Genes, Dopamine, and Behavior

Animal models provide further clues. Mice with mutations in dopamine transporter genes show altered dopamine signaling and social behaviors. Other mouse models, such as those lacking the SHANK3 gene, display excessive dopamine release in new environments, along with repetitive behaviors. Interestingly, when these mice are placed in familiar settings, their behavior improves, suggesting that environmental enrichment could help reduce symptoms.

The Kynurenine Pathway: A Link Between Serotonin, Dopamine, and Inflammation

Another piece of the puzzle comes from the kynurenine pathway, the main route by which tryptophan (the building block of serotonin) is broken down. Inflammation in the body can push tryptophan metabolism toward kynurenine production. Kynurenine can then be converted into either kynurenic acid, which protects the brain, or quinolinic acid, which is toxic at high levels. An imbalance here may worsen brain inflammation, oxidative stress, and damage to neurons. This could affect both serotonin and dopamine systems.

Toward More Balanced Treatments

Given these findings, some researchers and clinicians are experimenting with new approaches. Instead of relying only on neuroleptics, they are looking at natural compounds that support balance between neurotransmitters. For example, 5-HTP (5-hydroxytryptophan) can boost serotonin production. Branched-chain amino acids (BCAAs) may reduce excess dopamine by competing with its building blocks. Taurine, another amino acid, may help calm neuronal activity and protect against glutamate excitotoxicity.

A Puzzle with Many Pieces

The story of serotonin, dopamine, and autism is far from complete. Both neurotransmitters clearly play roles, but they interact with many other systems, including inflammation, genetics, and environment. Autism is a highly diverse condition, which makes finding universal treatments difficult. Still, the more researchers learn about brain chemistry, the closer they get to therapies that can improve quality of life, reduce aggressive or repetitive behaviors, and enhance social connection.

For now, the lesson is clear: the brain's balance between serotonin and dopamine is delicate. When disrupted, it can contribute to the behaviors and challenges seen in autism. By continuing to explore these pathways, science hopes to uncover targeted, effective, and safer ways to support people on the spectrum.

Autism and the Kynurenine Pathway

In recent years, scientists have paid more attention to the kynurenine pathway (KP), a major route for breaking down the amino acid tryptophan. This pathway produces several chemicals that can either protect or harm the brain. For people with autism spectrum disorder (ASD), studies suggest that this pathway often works abnormally.

Research has shown that children with ASD may have higher levels of quinolinic acid (QUIN), a compound that can damage brain cells, and lower levels of kynurenic acid (KYNA), which normally protects them. For example, Lim and colleagues (2016) found increased levels of kynurenine and QUIN in autistic children, while other teams reported reduced KYNA in their blood and urine. In adults with autism, recent studies also confirmed low KYNA and tryptophan, pointing to a long-lasting imbalance. These chemical changes are often linked to repetitive behaviors and persistent inflammation, both common in ASD.

Inflammation seems to be a key driver of this imbalance. Pro-inflammatory molecules like interleukin-6 (IL-6) activate enzymes that push tryptophan into the harmful side of the pathway, raising QUIN levels. Animal studies show similar effects when the immune system is challenged by bacteria or viruses. This close link between inflammation and metabolism may explain why many autistic individuals also show signs of chronic immune activation.

These findings also open the door to possible treatments. Blocking the production of QUIN, boosting KYNA, or controlling inflammation through specific drugs could restore balance in the brain. Supplementing KYNA or its analogs may further strengthen natural protection. Understanding how the kynurenine pathway goes wrong in autism could help scientists design new therapies that improve brain health and reduce symptoms.

PANS and PANDAS: When the Immune System Shapes the Mind

In recent years, doctors and researchers have paid closer attention to two puzzling pediatric conditions: PANS (Pediatric Acute-onset Neuropsychiatric Syndrome) and PANDAS (Pediatric Autoimmune Neuropsychiatric Disorder Associated with Streptococcal infections). These disorders strike suddenly, changing a child's life almost overnight. Once healthy children may wake up showing severe obsessive-compulsive behaviors, anxiety, eating restrictions, or tics. Families often describe the shift as "like a switch flipped," leaving them searching for answers.

PANS covers a broad range of triggers. These include infections such as Mycoplasma pneumoniae, influenza, sinusitis, and group A streptococcus (GAS), but also non-infectious stressors. PANDAS, however, is a narrower subgroup, directly tied to strep infections. In these cases, the sudden onset of symptoms follows closely after a streptococcal infection, such as strep throat. Children with PANDAS often experience a relapsing-remitting pattern, where symptoms flare with each new infection and fade with recovery.

Diagnosing these conditions requires careful clinical evaluation. Doctors look for the abrupt start of obsessive-compulsive disorder (OCD) or restrictive eating, combined with other psychiatric or neurological symptoms. These may include extreme anxiety, depression, irritability, regression in behavior, school decline, tics, or even physical issues like bedwetting and sleep problems. For PANDAS specifically, symptoms must begin between ages 3 and puberty and show a clear link to strep infections. Unlike other psychiatric conditions, the timeline and dramatic presentation are key clues.

Scientists believe the immune system is at the heart of both syndromes. Normally, the brain is protected from the immune system by the blood-brain barrier. But in PANS and PANDAS, antibodies may cross this barrier and mistakenly attack brain tissue, particularly the basal ganglia–regions that control movement, emotions, and behavior. This idea echoes what is seen in Sydenham's chorea, a neurological disorder linked to rheumatic fever, where strep antibodies target the brain and cause abnormal movements. In PANS and PANDAS, a similar process may trigger OCD, tics, and emotional instability.

Treatment strategies reflect this immune connection. First, if an infection is detected, antibiotics such as penicillin or amoxicillin are prescribed. In some children with recurrent episodes, long-term antibiotic prophylaxis is considered. When symptoms are severe, doctors may use corticosteroids to calm inflammation or intravenous immunoglobulins (IVIG) to neutralize harmful antibodies. In the most difficult cases, plasma exchange has been tried to directly remove these antibodies from the bloodstream.

Beyond immune therapies, children also need psychological and educational support. Cognitive-behavioral therapy (CBT) helps them manage OCD and anxiety, while medications like selective serotonin reuptake inhibitors (SSRIs) may provide additional relief. Schools often adapt by reducing workloads or offering exam accommodations. Families are also encouraged to explore complementary approaches, including stress management, good nutrition, and physical activity.

Researchers have also developed diagnostic tools such as the Cunningham Panel, which measures antibodies that may attack brain proteins, including dopamine receptors and tubulin. While its clinical value is debated, it highlights growing efforts to uncover the biological roots of these syndromes.

PANS and PANDAS remain controversial, with ongoing debates about their exact mechanisms and treatment standards. Still, awareness is rising. What is clear is that early recognition and intervention can dramatically improve outcomes. By bridging pediatrics, neurology, psychiatry, and immunology, medicine is slowly piecing together how infections and immunity may shape the developing brain. For families, this knowledge brings both hope and a roadmap toward recovery.

Autism and Aggression: Understanding the Hidden Triggers

Aggression and violent behaviors in people with autism remain one of the most difficult challenges for families and professionals. These behaviors can block social development and education, often leaving families confused and exhausted. Scientists have shown that aggression in autism does not come from a single source, but from a mix of brain chemistry, immunity, hormones, and even nutrition.

Neurotransmitter Imbalance: Dopamine and Serotonin

Dopamine, the brain chemical tied to motivation and learning, becomes a double-edged sword when overactive. Too much dopamine leads to impulsivity and agitation. Some strategies, such as using branched-chain amino acids (BCAA), try to reduce dopamine levels by limiting the raw materials the brain uses to make it. On the other side, serotonin plays a calming role in controlling impulses. When serotonin is too low, the brain struggles to regulate anger and aggression. Selective serotonin reuptake inhibitors (SSRIs) and precursors like tryptophan may help, though tryptophan must be carefully balanced to avoid overstimulation of NMDA receptors.

Temporal Lobes and Glutamate

Another clue comes from the temporal lobes of the brain. Problems in blood flow here, often linked to inflammation, can raise levels of glutamate, a powerful excitatory chemical. This buildup of glutamate is tied to sudden bursts of violence. Anti-epileptic drugs like oxcarbazepine can calm neural activity and reduce aggression without heavy side effects.

Stress, Noradrenaline, and the Fight-or-Flight System

Stress is another driver. The body's stress system, called the HPA axis, increases noradrenaline levels. This chemical primes the body for "fight or flight," but in autism it can push children into explosive reactions. Off-label drugs such as guanfacine and propranolol can reduce this hyperactive stress response and ease aggressive outbursts.

Immune and Hormonal Factors

The immune system also plays a part. Thyroid problems, especially autoimmune hypothyroidism, are more common in autistic individuals. They can overstimulate the stress system and raise cortisol, the "stress hormone," which is linked to aggression. Nutritional support, like selenium, iron, and vitamins A and E, can help stabilize thyroid function. Hormones such as testosterone also influence aggression. High testosterone in autistic boys has been tied to violent behavior, and some drugs like spironolactone have shown promise, though side effects limit their use.

Gut, Diet, and Aggression

The gut-brain connection is another surprising piece. Sensitivity to gluten and casein proteins can worsen aggression, especially when the gut barrier is "leaky." A gluten- and casein-free diet often improves behavior. Gut infections and imbalances, like overgrowth of Clostridium difficile, can also trigger violence. Targeted antibiotics plus probiotics such as Lactobacillus rhamnosus GG may restore balance.

The "Awakening" Phase

Beyond biology, some children experience what parents call the "awakening phase." After infections or toxins are treated, the child's brain seems to "wake up." Families see progress but also face a wave of anxiety and violent behaviors. This stage is painful and confusing: parents may think their child is regressing, but in fact, the brain is revealing hidden imbalances that were masked before. This phase requires careful management—emotional support, anxiety treatment, and patience. Though exhausting, it can mark an important step toward long-term progress.

Aggression in autism is not simple misbehavior. It is a biological storm shaped by brain chemistry, immunity, stress, diet, and hormones. Understanding and treating these roots can transform life for autistic individuals and their families.

Autism and Language: Cracking the Communication Code

Language difficulties are often the first red flag that makes families and health professionals suspect autism spectrum disorder (ASD). These challenges can vary widely. Some children do not speak at all, while others show subtle but real struggles with the social use of language. In many cases, language development is delayed, and in others, children lose skills they once had, a phenomenon called language regression. This loss of abilities highlights how complex the brain and immune processes behind language really are.

Science is uncovering surprising players in this story. Mast cells, usually known for their role in allergic reactions, are now linked to brain function. In autistic individuals, these immune cells can become overactive, sparking chronic inflammation in brain regions tied to speech and listening, such as the prefrontal cortex. This inflammation disrupts the brain's flexibility, known as plasticity, which is vital for building and adapting language skills. At the same time, the body's endocannabinoid system – involving cannabinoid receptors like CB1 and CB2 – appears to play a major role in regulating brain inflammation and communication. If this system is off balance, it may contribute to speech problems and other key autism symptoms.

Another piece of the puzzle is hyperarousal, or a state of constant overactivation. Many autistic people experience hypersensitivity to sound, light, or touch. This overload can block the brain circuits needed for learning language and holding conversations. Research has shown that propranolol, a common blood pressure medicine, can help calm this physiological storm. By reducing the body's stress response, it creates a more stable emotional state that allows for better focus, reciprocity in conversation, and clearer language processing.

Nutrition and targeted therapies are also entering the spotlight. Omega-3 fatty acids, well known for supporting brain health, have been linked to improvements in word use and gesture-language combinations in young children with autism. Folinic acid and methylcobalamin, two nutrients that act on folate metabolism and oxidative stress, have shown measurable effects on both expressive and receptive language. Tetrahydrobiopterin (THB), a cofactor needed to produce dopamine and serotonin, has been connected to gains in both verbal ability and social interaction.

These insights underline a central message: language problems in autism are not just isolated symptoms. They reflect a complex interplay between the immune system, brain chemistry, and sensory responses. Innovative therapies that reduce inflammation, regulate hyperarousal, and strengthen brain communication pathways are offering new hope.

Among all these interventions, folinic acid supplementation is drawing special attention. As an active form of folate, it plays a vital role in the brain's development, helping make neurotransmitters and regenerating nerve cells. Children with genetic mutations such as MTHFR or folate transport issues benefit the most. Studies have shown that folinic acid treatment can improve not only language but also social interaction, while reducing repetitive behaviors and hyperactivity.

For families and clinicians, this research paints a hopeful picture. By combining medical science, nutrition, and education, it may be possible to unlock language potential in many children with autism, giving them a stronger voice in both learning and life.

Beyond Every Protocol: Understanding Autism's Complexity

For decades, brilliant doctors have tried to decode autism, searching for theories that could explain it and protocols that might treat it. Many of these approaches have offered hope. Yet the danger lies in believing autism can be understood or treated through a single lens. Too often, a narrow focus blinds us to the bigger picture, and children lose therapeutic opportunities that could have helped them.

Autism is not one-size-fits-all. Each child is unique, and their responses to treatments vary widely. That is why doctors must stay open to interdisciplinary approaches. Integrating nutrition, neurology, immunology, psychology, and genetics is not just wise; it is necessary. Only then can therapies be personalized to meet the specific needs of each child. Treating autism with a rigid "protocol" risks ignoring the individuality of those who live with it.

In the United States especially, the medical system is often shaped by money. Some doctors, trapped by their own success, promote therapeutic protocols that gather passionate followers. On social media, a single success story may be exaggerated until it becomes dogma. Parents desperate for hope may cling to these "miracle cures," forgetting that autism is not a uniform condition.

Still, the work of these doctors deserves recognition. Each theory adds something to our collective understanding, even if no single one can capture the full complexity of autism. The reality is that neuroinflammation, gut dysfunction, immune dysregulation, mitochondrial weakness, and genetic differences all intertwine. No one pathway explains everything. That is why long-term observation is vital. A therapy that looks promising for three months may fail in three years. Only by tracking outcomes over time can doctors truly separate short-term hype from sustainable benefit.

Here are some of the most notable protocols and the doctors behind them. Their work has shaped the autism conversation for decades.

Dr. Daniel Rossignol, Dr. Jerry Kartzinel, Dr. James Neubrander, and Dr. Anju Usman in the U.S. argue that autism results from mitochondrial dysfunction, immune problems, encephalopathy, gut disease, heavy metal toxicity, and genetics. Their treatments include gluten-free and casein-free diets, methylcobalamin injections, antifungals, antibiotics, hyperbaric oxygen therapy, anti-epileptics, Alzheimer's medications for cognition, and chelation therapy to remove heavy metals.

Amy Yasko and Ben Lynch emphasize the intersection of genetics and epigenetics. They recommend cofactors like B12, B6, folates, zinc, and magnesium to bypass methylation and transsulfuration pathway issues.

Dr. Michael Chez suggests strokes, neonatal trauma, encephalopathies, epilepsy, and epileptiform activity explain autism. His treatments include valproic acid for EEG anomalies and speech delays, memantine and carnosine for language, and even stem cell therapy.

Dr. Suzanne Goh highlights mitochondrial deficits, strokes, brain trauma, and epilepsy. She prescribes valproic acid, supplements like carnitine and CoQ10, and behavioral therapies such as ABA and music therapy.

Dr. Robert Naviaux views autism as mitochondrial dysfunction coupled with a chronic "cell danger response." His controversial therapy is intravenous suramin.

Dr. Theoharis Theoharides and Dr. Nicola Antonucci focus on allergies, mast cell activation, and microglial activation. They recommend luteolin, quercetin, palmitoylethanolamide, and Rerum to reduce inflammation in the brain.

Dr. Rosario Trifiletti links autism to autoimmunity triggered by strep or viral infections. His toolkit includes ibuprofen, antibiotics like amoxicillin-clavulanate, and IVIG for chronic immune phases.

Dr. Michael Goldberg proposes a viral origin, often herpesviruses. He prescribes antivirals like valacyclovir, antifungals, and SSRIs to improve cerebral blood flow.

Dr. Ty Vincent sees autism as immune system misregulation, advocating for low-dose immunotherapy.

Dr. Arthur Krigsman believes autism stems from intestinal inflammation, using corticosteroids, antibiotics, and antifungals.

Dr. Patrick Nemechek focuses on bacterial overgrowth in the small intestine. His protocol features high-dose omega-3s, inulin or rifaximin, and extra-virgin olive oil.

Dr. Natasha Campbell-McBride developed the GAPS diet, excluding grains, legumes, gluten, dairy, and refined carbs, paired with probiotics.

Dr. Yehezkel Ben-Ari of France blames disrupted GABA inhibition due to excess intracellular chloride. His approach uses bumetanide, a diuretic, to restore balance.

Each of these doctors has made important contributions. But taken alone, none of their protocols captures autism's full complexity. To impose one method as the answer is to underestimate the interplay between genetics, epigenetics, immunity, the gut, and the brain. Autism is not a monolith. It is a spectrum of diverse biological realities.

That is why personalization is key. A child with mitochondrial weakness may benefit from CoQ10, while another with immune dysfunction may need IVIG. A third might respond to dietary changes. But no single therapy works for all. Ignoring this diversity risks wasting precious time, money, and hope.

Even more, protocols must be tested long term. Families often report short-term improvements, only to see progress stall or reverse. Real science demands patience, observation, and adaptation.

In the end, the true message is simple: autism cannot be reduced to one theory or one cure. It demands humility from doctors, patience from families, and collaboration across disciplines.

And when politicians, like US President Donald Trump, suggest that folinic acid (leucovorin) might be a "miracle cure," it oversimplifies a condition far too complex for slogans. Hope must always be balanced with realism.

Mega-Analysis and Clinical Research on Autism

Autism spectrum disorder (ASD) is one of the most complex conditions in modern medicine, influenced by both genetic and epigenetic factors. Researchers now agree that clinical studies must move toward more precise and personalized approaches, focusing on how different children respond to different treatments.

For decades, many clinical trials have struggled to deliver conclusive results. Even though randomized controlled trials (RCTs) are considered the gold standard, most of those done in autism have used very small samples and treated autistic individuals as if they formed a homogeneous group. This has limited the value of the findings. The diversity of autism requires studies that reflect its complexity and individuality.

Around the world, researchers are developing and testing new therapies, from medications to behavioral interventions. The challenge is not a lack of ideas, but the speed at which they can be tested and validated. Technology offers new opportunities to accelerate this process, through global networking, large-scale data collection, and better collaboration between clinicians.

One of the key needs in autism research is agreement on outcome measures. Clinical studies should not only look at major improvements but also at small, early changes that can indicate whether a therapy is helping. These subtle improvements are often ignored, even though they may reflect real progress in a child's daily life. Larger sample sizes are also needed, and this is only possible through international collaboration in multicenter trials.

Another promising method is the n-of-1 trial. This is a clinical trial conducted with a single patient, using repeated cross-over designs in a randomized and blinded way. In 2000, the Journal of the American Medical Association described n-of-1 trials as providing the highest level of medical evidence. In autism, where each case is unique, this model could be very valuable. Imagine thousands of doctors worldwide conducting n-of-1 trials and entering their results into a shared database. Patterns would quickly emerge, showing which therapies work best for different types of children. The only condition is that these trials must follow strict ethical rules, such as those in the Helsinki Declaration of 1964.

Recent advances in big data and artificial intelligence (AI) make this vision possible. Instead of relying only on classical meta-analyses, which average out results and often lose important details, researchers could perform mega-analyses. These studies would use huge datasets from all over the world, capturing the complexity of autism more accurately. AI is particularly powerful in this field. It can analyze data in real time, track a child's response to treatment, and recommend the most promising strategies. It can also identify potential new therapies through drug repurposing, where existing medications are tested for new uses. This process is faster and cheaper than creating new drugs from scratch, and it could bring real hope for families.

The future of autism research will depend on international cooperation, advanced analytics, and personalized approaches. The combination of mega-analysis, n-of-1 trials, and artificial intelligence promises to reshape the way therapies are tested and delivered. For families facing the daily challenges of autism, this shift could mean faster access to effective treatments and a more hopeful future.

Translational Medicine: A Therapeutic Hope for Autism

Translational medicine is an emerging medical discipline that represents a real shift in how science connects to patient care. Often described as moving 'from bench to bedside,' it bridges basic laboratory research with real-world treatments for patients. In simple terms, it turns discoveries made in biology, genetics, or pharmacology into therapies that can directly benefit people.

Autism, a complex neurodevelopmental condition, highlights why translational medicine is so necessary. Traditional approaches in clinical research often struggle to keep up with autism's complexity. Randomized controlled trials, while useful, usually involve small samples and treat autism as if it were the same in every child. This one-size-fits-all model often fails, because autism is deeply heterogeneous—what works for one patient may not work for another.

The strength of translational medicine lies in its ability to connect scientists, clinicians, and families in real time. Instead of working in isolated silos, it encourages open data and open innovation. This means researchers and doctors can learn from each other, share information globally, and accelerate progress. Families dealing with autism cannot wait for decades of slow, bureaucratic research—new therapies need to be tested, adapted, and applied quickly.

Evidence-based medicine (EBM) has long been the gold standard in healthcare. But for autism, its rigid framework can sometimes slow down innovation. The reality is that only a subset of patients responds to any given treatment, and no single therapy is likely to help everyone. Translational medicine embraces this diversity by focusing on personalized care. It aims to identify biological subgroups within autism and match them with targeted treatments. This approach is more realistic than trying to design universal solutions.

Another challenge lies in the disconnect between research and practice. Many clinical observations never reach basic scientists, meaning valuable insights are lost. Translational medicine solves this by creating a two-way dialogue: laboratory findings inform clinical practice, and patient outcomes inform laboratory research. Together, this cycle speeds up discovery and ensures treatments are grounded in both science and reality.

The urgency is clear. While scientific tools have become more powerful and cheaper, the time it takes to bring new drugs to market is now much longer than it was in the 1950s. For conditions like autism, which affects millions worldwide, this delay is unacceptable. Meanwhile, mainstream care still relies heavily on behavioral therapies and medications such as risperidone, aripiprazole, or methylphenidate. These drugs may reduce aggression or hyperactivity but often come with serious side effects—weight gain, metabolic problems, sleep disturbances, or even neurological risks. Worse, they do not address the root causes of autism.

Some pioneering doctors are already practicing translational medicine on the ground, often with limited resources. By combining clinical insights with cutting-edge science, they are offering autistic children new opportunities for progress. These efforts show what is possible when innovation meets compassion. To truly change outcomes, translational medicine must be encouraged, supported, and expanded globally.

In the end, translational medicine is more than a research strategy—it is a lifeline. By breaking down barriers between the lab and the clinic, it offers hope that autism treatment can evolve beyond trial and error. For families, it is a reminder that science is not just about theories but about improving real lives, one child at a time.

Preventing Autism Is Possible

In recent years, growing research has highlighted the role of environmental factors in autism, with vitamin D deficiency emerging as one of the most significant. Many scientists now consider this deficiency a major risk factor that may contribute to the rising prevalence of autism spectrum disorders (ASD). As a result, systematic vitamin D3 supplementation during pregnancy is being recommended by some specialists. Doses between 5,000 and 10,000 IU per day are suggested as a rational target, not only to ensure proper fetal development but also to protect maternal health. Adequate vitamin D levels support the immune system, reduce inflammation, and are linked to a lower risk of complications during pregnancy. Since vitamin D can influence up to 10 percent of gene expression, its deficiency represents a true public health issue.

Another key factor concerns folic acid supplementation. For decades, the World Health Organization has advised pregnant women to take 400 micrograms daily to prevent neural tube defects such as spina bifida. However, recent findings show that the synthetic form of folic acid currently used is not always metabolized properly, especially in individuals with mutations of the MTHFR enzyme. Between 10 and 40 percent of the population may carry such mutations, making them unable to convert synthetic folic acid into its active form, methylfolate. This issue can disrupt vitamin B12 metabolism, as well as the methionine and transsulfuration cycles, which are essential for healthy neurodevelopment. Experts now recommend replacing synthetic folic acid with methylfolate and encouraging the consumption of natural food sources such as leafy greens, legumes, and animal liver.

The way babies are delivered also plays a role. The World Health Organization recommends a cesarean rate of about 15 percent, yet countries such as Morocco report levels as high as 61 percent. Babies born through natural delivery receive their first exposure to their mother's vaginal microbiome, which helps build a healthy gut flora. This flora is crucial for immune and neurological development. Too many C-sections may deprive children of this essential start. Likewise, breastfeeding plays a protective role in establishing a strong intestinal microbiome. When breastfeeding is not possible, infant formulas should be enriched with protective compounds like 2'-fucosyllactose to support gut health.

Vaccination remains one of the most important public health achievements and continues to save millions of lives. However, some experts caution that giving vaccines to children with severe vitamin D deficiency may increase the risk of inflammatory or autoimmune reactions. Supplementing with vitamin D could reduce such risks and improve outcomes. Another concern is the widespread use of acetaminophen (paracetamol) in children. Even at normal doses, acetaminophen can deplete glutathione and sulfate reserves, harming mitochondrial function. Using this drug preventively to manage post-vaccine fever should be avoided. Antibiotics, too, must be prescribed carefully, since they can disrupt gut flora balance. Each antibiotic course should be accompanied by protective measures, including probiotics, to maintain microbiome stability.

Environmental toxins are another piece of the puzzle. Heavy metals and pesticides are strongly linked to neurological and developmental disorders. Policies promoting sustainable agriculture, pollution reduction, and healthier environments are essential for long-term prevention strategies. In addition, thyroid dysfunction in women of childbearing age should be screened and managed early, as maternal hypothyroidism is a known risk factor for neurodevelopmental issues. Obesity, smoking, and air pollution, all associated with lower vitamin D levels, further increase autism risk.

Prevention also depends on early detection. Today, autism is a major public health concern. The ability to diagnose it should not be limited to specialized centers. Pediatricians, general practitioners, and educators must all be trained to recognize early warning signs. Earlier intervention leads to better outcomes, and reducing delays in diagnosis could transform lives. Research shows that autism is not a fixed genetic trait but a complex condition influenced by both environment and biology. With proper action, prevention is possible.

The fight against autism requires global cooperation. Under the leadership of the World Health Organization, countries must develop clear guidelines for vitamin D supplementation, rational use of medications, protection of the microbiome, and reduction of environmental risks. With political will and international collaboration, autism prevention could move from possibility to reality, giving future generations a healthier start in life.

Fecal Transplants and Autism: What the New Science—and New Rules—Really Say

In 2019, a small but widely discussed study in Scientific Reports reported that a special form of fecal microbiota transplantation (FMT) called Microbiota Transfer Therapy helped children with autism who also had serious gut problems. Researchers at Arizona State University adapted a protocol pioneered by Australian gastroenterologist Thomas Borody, a longtime champion of microbiome-based care. In phase one, 18 children received two weeks of oral vancomycin, a bowel cleanse, and then high-dose donor microbiota followed by lower-dose maintenance for 7–8 weeks alongside an acid-suppressing medicine. After 10 weeks and an eight-week follow-up, gastrointestinal symptoms fell by about 80%, and autism-related measures improved. Two years later, phase two follow-up suggested durable benefits: ongoing GI improvement and a large drop in autism severity scores, with some participants no longer meeting spectrum criteria. Those findings didn't prove causation or work for everyone, but they helped push the idea that the gut–brain axis matters in autism.

What actually is FMT? At its simplest, FMT moves stool from a carefully screened donor into a patient's gut, aiming to restore a disrupted microbiome. Versions of the idea go back centuries in Chinese medical texts, while modern case reports began in the 1950s for life-threatening C. difficile colitis. Today, FMT (or newer microbiome products) is best established for recurrent C. difficile infection that doesn't respond to standard antibiotics, and that's where U.S. regulation has changed the most in the past few years.

The new U.S. legal landscape

Until recently, FMT lived in a regulatory gray zone. The Food and Drug Administration (FDA) used "enforcement discretion," allowing doctors to perform FMT for recurrent C. difficile without filing an Investigational New Drug (IND) application—as long as patients gave specific informed consent. In late 2022, FDA finalized guidance that narrowed this discretion. The policy now generally requires an IND unless strict criteria are met, and it no longer extends enforcement discretion to stool sourced from stool banks due to safety concerns. In other words, off-the-shelf preparations from stool banks are not covered by the old policy; clinicians who still perform FMT under discretion must use rigorously screened donors arranged for individual patients and meet the consent requirements. At the same time, the "market" has shifted from traditional FMT toward regulated products. In November 2022, FDA approved REBYOTA (fecal microbiota, live–jslm), the first microbiota-based product for preventing recurrent C. difficile. In April 2023, FDA approved VOWST (fecal microbiota spores, live–brpk), the first oral microbiome therapy for the same indication. These approvals created a commercial, prescription-only path for microbiome restoration in C. difficile and signaled how the agency wants to handle this field—through standardized, quality-controlled biologics rather than ad-hoc transplants.

What this means for autism research

Because autism is heterogeneous and the evidence for FMT is still preliminary, FMT is not an FDA-approved treatment for autism. The Arizona State study was small, open-label, and focused on children with both autism and significant GI symptoms. It is encouraging but not definitive. The clear next steps are larger, randomized, controlled trials that include strict donor screening, standardized manufacturing, and long-term safety monitoring—exactly the direction FDA is nudging the field.

Where safety fits in

Safety is the reason the rules tightened. FDA has issued repeated safety alerts about the risk of transmitting multidrug-resistant organisms through FMT if donors are not screened properly. Approved products are manufactured under quality systems with validated pathogen testing, traceability, and lot control. For any off-label, experimental use (including autism), families should expect an IND-supervised protocol, detailed informed consent, and close follow-up.

The bigger picture: gut–brain links

Why did FMT help some children in that study? Researchers suspect several mechanisms: restoring beneficial species (such as Bifidobacterium and Prevotella) that are often depleted; increasing short-chain fatty acids that support the gut lining; and reducing microbial by-products that may influence brain signaling and immunity. Mouse studies have shown that transplanting microbiota from people with autism can transfer social and repetitive-like behaviors, while targeted metabolites can partly reverse them—hints that the microbiome is a meaningful modifier, not a magic bullet.

What to watch next

Expect more trials testing defined microbial consortia or spore-based capsules rather than raw stool, often combined with diet, prebiotics, or antibiotics. Some groups are also exploring whether microbiome therapy could lower GI pain, reduce irritability, or improve sleep—indirect paths that might help learning and behavior. If future studies show benefit, regulators will likely favor standardized live biotherapeutic products with clear labeling and post-market safety plans.

The bottom line

FMT has moved from fringe to framework—especially for C. difficile, where two FDA-approved options now exist. For autism, the science is promising but early. Families considering microbiome-based care should enroll in clinical trials when possible, ask about IND oversight and donor screening, and be cautious about unregulated stool products. Microbiome medicine is evolving fast, but the safest progress will come from rigorous, well-designed studies that balance hope with proof.

The Future Perspective: The Cell Danger Response Theory

Our cells constantly monitor their environment to detect potential threats. One way they do this is through purinergic signaling. Purines, beyond being essential parts of DNA and RNA, are also found in critical biomolecules such as ATP, GTP, cyclic AMP, NADP, SAM, and coenzyme A. Among them, ATP is the best known. Inside the cell, ATP acts as an energy source. But when found outside the cell, extracellular ATP serves as a "danger signal."

Cells monitor the levels of purines around them using special receptors called P1, P2X, and P2Y. These receptors detect changes in purine concentration and send messages to the cell, especially to the mitochondria, signaling that a potential danger is present. This system, called purinergic signaling, is part of the body's early warning defense system.

The Cell Danger Response (CDR) theory suggests that abnormalities in purinergic signaling can keep cells in a state of chronic danger response. This prolonged state may prevent normal healing and contribute to conditions like autism. The CDR is a natural and universal response to stress, injury, or infection. It is designed to protect the cell and trigger repair. But when the process gets "stuck," healing cannot progress normally. This disrupted cycle can alter how cells communicate with their environment and affect long-term brain development.

Some researchers compare the CDR to post-traumatic stress disorder. When triggered in early childhood, it may lead to autism or other chronic conditions. If activated later in life, it could contribute to illnesses such as chronic fatigue syndrome, Gulf War syndrome, or certain autoimmune diseases. In short, any stressor—such as viral infections, physical trauma, or even intense exercise—can trigger the CDR. Usually, the body resolves the process and returns to balance, often stronger than before. But in certain individuals with genetic or epigenetic vulnerabilities, the danger response persists abnormally.

One of the key figures behind this theory is Dr. Robert Naviaux, professor of medicine, pediatrics, and pathology at the University of California, San Diego. Since 2013, he has published groundbreaking research showing that the CDR can be reversed in animal models of autism. His work suggests that this mechanism may be a unifying explanation for many different causes of autism.

In 2017, Naviaux published results from a small phase I/II clinical trial using suramin, a century-old drug originally designed to treat African sleeping sickness. Ten children with autism participated. Five received low-dose suramin injections, while five received a placebo. The children treated with suramin showed remarkable improvements in language, social behavior, and repetitive movements. Families reported striking changes—some children spoke full sentences for the first time. However, the effects faded after about six weeks, showing that repeated treatment may be needed.

Suramin works partly by blocking extracellular ATP signaling, acting as an "antipurinergic" drug. The trial was small, but it demonstrated something revolutionary: a potential common mechanism underlying autism, regardless of genetic or environmental causes. The next step for researchers is to develop new antipurinergic drugs that might provide longer-lasting results with fewer side effects.

The CDR hypothesis proposes that autism, despite its many apparent causes, may share a single underlying cellular response. When cells perceive danger, they launch protective biochemical cascades. But if the switch to "healing mode" never occurs, communication between cells and neurons is disrupted. This could explain why so many different genetic and environmental risk factors all converge toward similar behavioral symptoms.

The promise of targeting the CDR is immense. Instead of waiting years for therapies to address root causes like gut imbalance, immune dysfunction, or metabolic errors, doctors could intervene earlier by calming the danger response. By reducing neuroinflammation and restoring brain function, children could make rapid progress, while clinicians work on long-term solutions.

In summary, the Cell Danger Response theory provides a unifying framework to understand autism as a disorder of blocked healing. Early studies like Naviaux's trial offer hope but also caution: much larger, controlled studies are required to confirm safety and effectiveness. If proven, antipurinergic therapies may one day transform autism care, helping children move past locked states of cellular alarm into pathways of true recovery.

Future Perspectives: Transcranial Magnetic Stimulation

In recent years, the rise of non-invasive brain stimulation techniques has sparked hope as potential treatments for the main symptoms of autism spectrum disorder (ASD).
Among these approaches, transcranial magnetic stimulation (TMS) has gained particular attention. This method uses electromagnetic fields to stimulate brain neurons.
It can be applied both to study brain function and to treat various neurological and psychiatric conditions.

TMS was first described as a method of brain modulation by Anthony Barker in 1985. It is based on the principles of magnetic induction that were first introduced by Michael Faraday in the 19th century.
The concept is straightforward: a short, high-intensity electrical current passes through a coil of wire placed on the scalp. This generates a brief magnetic field strong enough to depolarize neurons near the stimulation site, producing physiological and behavioral effects that vary depending on the targeted brain region, also known as the cortical target.

A single session of TMS is usually exploratory or experimental, while repetitive TMS (rTMS) can modify brain responses for much longer periods. The effects of rTMS can last for weeks or even months after a series of stimulations.
Repetitive TMS involves applying multiple pulses of magnetic stimulation to a brain region, with frequencies ranging from 0.5 Hz to 20 Hz. Low-frequency stimulation tends to suppress cortical excitability in the targeted area, while higher

frequencies, above 5 Hz, tend to enhance excitability. However, the results can vary greatly between individuals.

For more than 20 years, neuroscience laboratories and hospital departments specializing in neurophysiology, neurology, and psychiatry around the world have used TMS for various research and therapeutic purposes.
The method is generally well tolerated, but some side effects have been reported. This led to the creation of official usage guidelines, first by the U.S. National Institutes of Health (NIH) in 1996, and later expanded globally by the International Society for Transcranial Stimulation and the International Federation for Clinical Neurophysiology.
These recommendations helped establish TMS and rTMS as safe and acceptable procedures in clinical practice.

A recent meta-analysis reviewed 23 studies on the effects of rTMS in autism spectrum disorder, including 12 controlled trials. The analysis found significant reductions in repetitive behaviors.
However, the findings were limited by statistical biases, largely due to the small size of most study groups.
The effectiveness of rTMS depends on several factors, including stimulation frequency, the specific brain target, the number of sessions, their duration, and the intervals between stimulations.

While the potential of TMS is exciting, it also raises ethical concerns. Media coverage, especially in the United States, often portrays TMS as a miracle solution for autism, fueling optimism that is not yet justified by scientific evidence.
It is essential to conduct rigorous clinical trials to clarify how TMS works, test its true effectiveness, and identify the most appropriate brain targets for therapy.

We live in a time when bold promises are fueled by rapid progress in nanotechnology and neuroscience. Entrepreneurs and scientists worldwide claim they are close to "modulating" or even "enhancing" cognitive abilities with brain implants, such as those promoted by Elon Musk, or with non-invasive techniques like TMS.

But these developments must be surrounded by strong ethical safeguards. Despite the lack of regulatory approval for TMS in autism, commercial clinics are already offering treatments to parents of autistic children, often with exaggerated promises of spectacular results.

Science must always respect the principles of bioethics, especially when dealing with vulnerable populations like children.

Most published studies report no serious adverse effects from TMS. The side effects observed, such as headaches or mild discomfort, are usually temporary and well tolerated.

Nevertheless, more rigorous studies are necessary, with stricter protocols, double-blind designs, and longer follow-up periods to measure outcomes accurately.

Without these safeguards, it would be risky to recommend TMS as a reliable treatment for autism.

It is also important to recognize that TMS is unlikely to improve all the different aspects of autism. The condition is highly complex and varies widely from person to person.

Expecting one form of brain stimulation to address every symptom is unrealistic. For now, the evidence remains too limited to recommend TMS as a standard therapy for children with autism.

Still, researchers are optimistic. The technology is non-invasive, relatively safe, and shows some promise in reducing repetitive behaviors and improving social functioning.

But caution is needed. TMS should not be promoted as a cure, but rather as an experimental approach that requires more study. Only with larger, better-designed trials can the scientific community determine whether TMS has a true role to play in autism treatment.

For families desperate for solutions, the hope of a breakthrough is understandable. But until solid evidence emerges, experts stress the importance of patience and caution.

Children with autism deserve safe, effective, and ethically sound treatments–based not on hype or media promises, but on proven science.

Transcranial magnetic stimulation may one day become part of the autism therapy toolbox, but for now, it remains a technology full of potential that still requires careful testing and validation.

Stem Cell Therapy: Between Hope and Illusion

On the internet, dozens of websites and social media ads promise that stem cell therapy can "cure" autism. Clinics in countries with looser regulations–like Panama or Thailand–claim to have helped hundreds of children. Families, desperate for answers, often spend thousands or even tens of thousands of euros chasing this hope. Online groups reinforce these promises, encouraging parents to believe stem cells might be the miracle they've been waiting for.

Stem cells are powerful because they can renew themselves and transform into specialized cells. Scientists dream that one day they could treat many diseases. But the science is still young, and some providers are taking advantage of that gap. In 2019, the U.S. Food and Drug Administration (FDA) issued a strong warning against unproven stem cell therapies. It emphasized that only treatments officially approved or under strict clinical investigation are allowed–and currently, autism is not one of them.

The risks are not imaginary. Stem cells can sometimes change into the wrong type of tissue, fail to function, or even fuel tumor growth. Despite this, some doctors in the U.S. and abroad continue to sell the therapy. A small 2019 trial in Panama, for example, reported "safe use" of stem cells from umbilical cords, but the study lacked scientific rigor and raised ethical questions.

Countries with strong regulation forbid commercial offers. In Morocco, for instance, stem cells can only be used in hospital research under ethical oversight. For now, while stem cell therapy sparks hope, the evidence is too weak. The dream remains, but science still has much to prove.

Advocacy for Early Autism Diagnosis

In medicine, both inertia and overzealous treatment can trap doctors, but when it comes to autism, it is inertia in early detection that is the most worrying. Despite alarming epidemiological data showing rising prevalence, early screening for autism often comes too late. This delay reflects not just difficulty in identifying the disorder, but also a widespread lack of awareness among many frontline professionals–general practitioners, pediatricians, and educators.

The saying often repeated in French medical schools–"You only find what you are looking for, and you only look for what you know"–fits perfectly here. For decades, the image of autism taught to doctors was a child withdrawn into repetitive movements or even self-injury. When I studied medicine twenty years ago in Casablanca, autism was grouped under "childhood psychoses." Sadly, many doctors still hold onto this outdated image.

Autism today is recognized as a spectrum. Every child is different, with unique traits and biological complexity. Comorbidities add to the challenge, and the disorder's unpredictable, non-linear progression makes detection even harder. To complicate matters, no routine biological marker exists, forcing clinicians to rely on observation and differential diagnoses. Without ongoing training, many frontline doctors miss the subtle early signs. Overworked and pressed for time, they sometimes reassure parents with the outdated phrase: "Your child is just developing at their own pace." But given today's alarming prevalence, such reassurances are no longer acceptable.

The average age of autism diagnosis still hovers around four to five years. This is far too late, since early intervention can dramatically improve outcomes. Some health systems even delay diagnosis in the name of multidisciplinary assessment, leaving families lost as their children receive shifting labels—"non-specific pervasive developmental disorder," "autism spectrum disorder," "Asperger's," "ADHD." But the exact label matters less than the speed of intervention. What matters most is rapid medical and educational support aimed at inclusion.

For years, families have faced a system that leaves them abandoned. The epidemiological reality demands urgent action. Early screening is not just a nice idea—it is a public health necessity. Blocking inflammation and disrupted signaling pathways early, especially in regressive autism, helps restore synaptic pruning. The earlier this happens, the more chance the brain's plasticity can close the developmental gap with peers. Every month lost is a missed opportunity.

Several screening tools already exist and are easy to use: the Ages and Stages Questionnaire (ASQ-3), the Modified Checklist for Autism in Toddlers (M-CHAT-R/F), the Screening Tool for Autism in Toddlers (STAT), the Social Communication Questionnaire (SCQ), the Social Responsiveness Scale, Vineland Adaptive Behavior, the Differential Ability Scale, and the Mullen Scales of Early Learning. These are simple, do not require deep training, and could be integrated into primary care. Diagnostic tools such as ADOS-2, ADI-R, CARS-2, and GARS-3, while more advanced, could also be part of every clinician's toolkit with minimal extra training. Ideally, medical schools should include these tools in their curriculum, while screening methods should also be shared with parents, since they are often the first to notice subtle signs.

Autism is not rare anymore; it is an unfolding pandemic. Some experts even warn its long-term impact on society could rival or surpass the consequences of Covid-19. Yet decision-makers seem oddly unconcerned, despite growing prevalence and delayed care. This inertia comes at a high cost for families and communities.

Scientific advances show why early detection matters. Autism is not simply a fixed genetic condition but a disorder with inflammatory and systemic features. Neuroinflammation disrupts synaptic pruning, blocking the brain's ability to refine its connections. If targeted anti-inflammatory interventions begin early, the pruning process can be restored, giving children a chance to recover developmental ground. Clinging to outdated diagnostic rituals in search of precision, while ignoring the urgency of early action, is irrational. We are not treating a label; we are rescuing a blocked neurodevelopmental process. Every minute counts.

My experience over the past nine years, combined with countless discussions with colleagues worldwide, supports this biological view. Autism is not an immutable label. Early detection and targeted therapy can halt the damage while brain plasticity still allows recovery. Once, autism was seen as a permanent state. But research on neuroplasticity proves otherwise. The brain, far from being static, can rewire itself, forming new connections in response to environment and therapy. Neuroinflammation, modifiable by treatment, plays a crucial role in how autism develops. In this sense, autism is dynamic, not fixed.

Today, the challenge is to close the gap between what science knows and what practice delivers. Screening must no longer be confined to specialized, multidisciplinary centers. It must become a shared skill across frontline healthcare—general practitioners, pediatricians, and even educators. With proper tools and training, subtle signs can be caught early, and interventions launched before the developmental gap widens. Families must no longer be left alone in uncertainty, waiting years for answers.

The instruments exist, the knowledge is advancing, and the stakes are clear. What is missing is political will and systemic change. Autism is a public health emergency, and every early diagnosis represents a chance for a child to thrive. The time for inertia has passed. Early screening and intervention must become the norm, not the exception. Only then can we meet this epidemic head-on and change the future for millions of children worldwide.

For a Truly Inclusive School

Italy has become a leading example when it comes to the inclusion of children with autism in schools.
The country's journey began in the 1970s, when two landmark laws (No. 118/1971 and No. 517/1977) guaranteed school integration for all children, regardless of disability. Since then, Italy has built one of the most inclusive school systems in the world, offering concrete solutions to the challenges of integrating children with autism.

One of the major strengths of the Italian model is its multidisciplinary network. This network brings together teachers, teaching assistants, school administrators, healthcare professionals, and doctors. Families are also considered part of this alliance. Together, these groups create a coordinated approach that ensures continuity and stability in the child's learning journey. Rather than focusing on academic skills alone, the Italian model is built on the concept of a "life project," where education is designed to prepare the child not only for school but for adulthood and society.

A key factor behind this success is the presence of a specialized educator in every classroom where a child with special needs is enrolled. These specialized teachers work side by side with regular teachers, playing the role of facilitator in the inclusion process. This partnership has transformed teaching strategies in several ways:

1. **Sharing knowledge**: Regular teachers learn to see each child as unique, with individual strengths, preferences, and

difficulties. They are encouraged to move beyond stereotypes and rigid categories.

2. **Co-planning**: By law, teachers in Italy are required to prepare lessons and projects together with specialized educators. This co-planning ensures that lessons are adapted to the needs of all students, not just those with special needs. It also encourages creativity and flexibility in teaching methods.

3. **Co-teaching**: Both the regular teacher and the specialized educator share responsibility for teaching the whole class. This joint approach avoids isolating the child with autism and ensures that inclusion benefits all students by promoting diversity and collaboration.

The Italian system's success is striking when compared to the French model, which has struggled with true inclusion.
While France has made progress since the 1990s, gaps remain. Autistic children are often included in preschool classes, but as they grow older, the system directs many of them into specialized units or institutions such as medico-educational institutes (IME). Instead of promoting inclusion, these pathways often separate autistic children from their peers, making integration more difficult.

French schools, despite political promises, continue to rely on outdated approaches. Inclusion is often seen as providing extra support or resources for the child, without changing the structure of the system itself. Many teachers are not adequately trained to understand autism. Old psychoanalytic theories, which are now discredited, still influence how autism is perceived in education. As a result, children are sometimes misunderstood or stigmatized instead of being
supported.

This situation is particularly relevant in Morocco, where the French model is still considered a reference. In reality, despite official speeches about inclusive schools, most autistic children in Morocco remain excluded from the classroom.
Some parents manage to win hard-fought battles to include their children, sometimes even through legal action. But these cases are exceptions. For the vast majority, the doors of education remain closed, leaving an entire generation behind.

For Morocco, it has become urgent to look beyond the French system and adopt lessons from the Italian experience.
Inclusion must not be reduced to slogans. It requires practical measures: training teachers, creating networks of professionals, and guaranteeing that every child with special needs has access to a specialized educator in the classroom.

The discoveries of brain plasticity reinforce the importance of such efforts. Neuroscience has shown that the brain has a remarkable ability to reorganize and adapt in response to experiences and interventions. For children with autism, this means that their potential is far greater than what outdated stereotypes suggest. If given the right support, many can make significant progress—sometimes far beyond what was once thought possible.

Setting low expectations for autistic children is not just a mistake; it is a denial of their fundamental rights. When educators or policymakers limit goals based on rigid assumptions, they rob children of opportunities that could have changed their lives. Denying a child access to appropriate education and inclusion is more than neglect—it is a serious injustice.

The Italian model proves that another way is possible. By embedding specialized educators in classrooms, encouraging teamwork between teachers, and involving families, Italy has built a system that respects every child's individuality. Rather than treating autism as an obstacle, it is seen as part of the natural diversity of human development.

For Morocco, the message is clear. It is time to abandon outdated frameworks and invest in inclusive education rooted in scientific knowledge and human rights. The challenge is enormous, but the cost of inaction is even greater. Every year of delay means more children excluded, more families isolated, and more wasted potential.

True inclusion is not about charity; it is about justice and equality. It is about recognizing that autistic children, like all others, have the right to learn, to grow, and to belong. A school system that excludes them is not only failing them, it is failing society as a whole. By embracing inclusion, Morocco could take a historic step forward, giving hope to thousands of families and proving that education can indeed be a bridge to a better future.

Beyond Autism: Uncovering Hidden Metabolic and Genetic Disorders

In recent years, autism has increasingly been viewed not only as a neurodevelopmental disorder but also through a biological and medical lens. Yet, many clinical presentations resembling autism can arise from rare metabolic or genetic disorders. International researchers, including Dr. Clara van Karnebeek and Dr. Sylvia Stöckler from B.C. Children's Hospital in Vancouver, have identified more than 80 metabolic errors that can cause intellectual disabilities often accompanied by autistic-like signs. Their systematic review has become a key reference for pediatricians and neurologists seeking to refine diagnoses.

A Spectrum That Extends Beyond Autism

The idea that autism might be mimicked by other conditions challenges both families and physicians. For decades, autism was classified narrowly, often as a static psychiatric label. Today, the medical community acknowledges that it is essential to separate true autism spectrum disorder (ASD) from conditions producing overlapping behaviors. Many of these disorders involve metabolic dysfunctions that, when properly identified, can be treated with diet, supplements, or targeted therapies. This distinction is not trivial–it can mean the difference between a life of severe disability and one of improved functioning through timely intervention.

Examples of Treatable Conditions

Metabolic errors can affect amino acids, cholesterol pathways, energy production, neurotransmitters, and even vitamin-dependent processes. Phenylketonuria (PKU), one of the best-known examples, results from a deficiency in the enzyme phenylalanine hydroxylase. Without treatment, PKU causes intellectual disability and autistic-like features. Yet, with a strict low-phenylalanine diet and tyrosine supplementation, children can develop normally. Other conditions, such as creatine transporter deficiencies or urea cycle disorders, can be improved with creatine supplements, protein restrictions, or ammonia-reducing agents. These targeted therapies highlight the importance of precise diagnosis.

The Role of Whole Exome Sequencing

The introduction of Whole Exome Sequencing (WES) has revolutionized the ability to detect these rare conditions. By analyzing the coding regions of DNA, WES allows clinicians to pinpoint mutations linked to rare metabolic syndromes. For families, this means faster answers and avoidance of inappropriate treatments. In countries with advanced medical systems, WES is now part of the diagnostic arsenal for children showing neurological or developmental delays. For low- and middle-income countries, integrating such technologies remains a challenge, yet its potential to change outcomes is undeniable.

Challenges for Healthcare Systems

Despite advances, many healthcare systems remain unprepared. Training for general practitioners, pediatricians, and educators often still reflects outdated notions of autism. Subtle metabolic red flags can be missed, and families may spend years without answers. The lack of awareness results in delayed intervention during the critical window of brain plasticity. For policymakers, the issue raises questions about resource allocation, ethics, and equity in access to genetic testing.

Hope and Caution

Stories from around the world illustrate both the promise and the limitations of medical science. Some children initially diagnosed with autism have experienced remarkable improvements once their true metabolic disorder was identified and treated. Others, however, face lifelong challenges even with optimal care. Researchers caution against overstating the hope. Not all metabolic errors are reversible, and for many families, management rather than cure remains the goal. What is clear is that awareness and early detection can dramatically change the trajectory for a subset of children.

Conclusion

Autism is increasingly recognized as a complex, biologically influenced condition. Yet it is vital to remember that not all autistic-like behaviors stem from autism itself. Behind the label, there may lie treatable metabolic and genetic disorders. By embracing new tools like Whole Exome Sequencing and encouraging broader clinical awareness, the medical community has a chance to offer hope where none previously existed. In this light, autism should be seen not as a final verdict, but as a starting point for deeper investigation.

Towards a Comprehensive Theory: The Seven Breaches Model

Autism remains, to my eyes, one of the most complex and fascinating puzzles in modern medicine. To understand this condition, it is essential to connect genetic, environmental, and metabolic factors, all of which interact in subtle ways to shape atypical neurodevelopmental pathways. To explain this complexity in a clear and accessible way, I drew inspiration from the Swiss cheese model, first used in the field of safety and risk management. This model illustrates how several layers of protection, though solid in appearance, contain weaknesses or "holes." When these holes align in an unfortunate sequence, the result is a global failure. Transposed to autism, this model helps explore multiple vulnerabilities which, when combined, lead to significant neurodevelopmental alterations.

In this framework, each "slice" of the model represents a protective barrier or a biological mechanism, while the "breaches" symbolize specific weaknesses or risk factors. Some of these weaknesses, such as vitamin D deficiency, act at the earliest stages of development, while others, like immune dysregulation or intestinal dysbiosis, appear later and reinforce the fragility of the system as a whole.

First Breach: Vitamin D Deficiency

Since my earliest research, vitamin D has emerged as a central element in understanding autism. This neurosteroid plays a vital role in brain development, and its insufficiency during pregnancy or the first years of life can lead to profound consequences. A lack of vitamin D compromises neuronal protection by limiting the expression of genes crucial for synapse formation and neuronal survival. In addition, it fosters chronic inflammation by destabilizing immune responses and can cause epigenetic modifications that alter gene expression at critical developmental stages. I consider vitamin D supplementation to be an absolute priority in any preventive approach.

Second Breach: Intestinal Dysbiosis and Microbiota Dysfunction

The gut microbiota is a true conductor of immunity and neurotransmitter regulation, but it is often unbalanced in autistic children. This imbalance, known as dysbiosis, often involves an overabundance of harmful bacteria, such as Clostridia, which are capable of producing neuroactive toxins like p-cresol or propionic acid. These compounds interfere with neurotransmitter regulation, notably by raising dopamine to toxic levels. In parallel, the disruption of the intestinal barrier, often referred to as "leaky gut," allows inflammatory molecules to pass into the bloodstream, where they can directly affect the brain. These gut-brain axis disturbances form a crucial breach in the cascade of dysfunctions linked to autism.

Third Breach: Immune System Dysregulation

The role of the immune system in autism's origins is central. I have often observed that both prenatal and postnatal inflammations contribute significantly to the emergence of symptoms. For instance, maternal immune activation triggered by infection or inflammatory stress during pregnancy can disturb synapse formation in the fetus. In some cases, maternal autoantibodies cross the placental barrier and target specific neuronal antigens, altering brain development. These immune weaknesses, when added to other vulnerabilities, markedly increase the risk of neurodevelopmental disorders.

Fourth Breach: Neurotransmitter Imbalances

Chemical imbalances in neurotransmitters, particularly serotonin and dopamine, are frequent in autistic children. A striking paradox has been noted: high serotonin levels in the blood often coexist with a deficiency in the brain. This imbalance deeply affects social and emotional interactions. Dopamine, meanwhile, is often overproduced or poorly regulated, fueling repetitive behaviors and aggressiveness. These chemical anomalies, when combined with the other breaches, further disrupt neural circuits.

Fifth Breach: Metabolic and Mitochondrial Dysfunctions

Mitochondria, the cell's powerhouses, play a critical role in neuronal development. Their dysfunction—whether from insufficient energy production or excessive oxidative stress—weakens neurons during critical growth stages. In addition, toxic compounds produced by a dysbiotic microbiota, such as p-cresol, can disrupt energy metabolism by reducing essential cofactors like coenzyme A. These metabolic deficits, often invisible, create an underlying structural vulnerability.

Sixth Breach: Environmental Exposures and Epigenetic Factors

Gene-environment interactions are unavoidable in explaining autism. Prenatal exposures to toxins such as pesticides or heavy metals, when combined with genetic predispositions, alter brain developmental trajectories. These influences interfere with essential metabolic pathways, such as DNA methylation, and hinder synaptic development. Such environmental exposures form insidious but decisive breaches.

Seventh Breach: Neuroinflammation and the Cellular Danger Response

Chronic brain inflammation is probably the most visible and studied breach. The cellular danger response (CDR)—a defense mechanism meant to protect the body against threats—becomes pathological when it persists too long. Microglia, the brain's immune cells, produce inflammatory cytokines that disrupt neural connectivity and the balance between excitation and inhibition. While these alterations were long seen as irreversible, recent discoveries about brain plasticity offer new therapeutic hope. Research suggests that even after prolonged neuroinflammation, targeted interventions can help restore some brain functions, opening promising perspectives for improving the quality of life of affected children.

The Seven Breaches model has allowed me to structure my understanding of autism and present a comprehensive vision. This condition does not stem from a single cause, but from the unfortunate alignment of multiple vulnerabilities that, when combined, reach a tipping point. For example, a mother suffering from vitamin D deficiency may give birth to a child with a preexisting genetic fragility. Prenatal exposure to environmental toxins or an unbalanced microbiota can then aggravate this fragility, triggering a cascade of metabolic and inflammatory disruptions. By identifying and correcting these breaches, I firmly believe we can reduce autism's impact on millions of children and families. This model provides a solid foundation to think differently about autism—with the hope of a brighter future.

Epilogue

This book does not claim to be a finished work. On the contrary, it is a step along a path I continue to walk with humility and determination. Autism, as I have explored it through research, clinical practice, and my own journey as a father, remains a vast and complex territory. Each chapter of this book reflects not only my certainties but also my doubts, my successes as well as my failures. It is a testimony in constant evolution, enriched by the children I have had the honor to accompany and by the families who entrusted me with their hopes.

This book is also a cry of hope. Hope that one day we will be able to unlock the mysteries of autism and provide solutions that are tailored, personalized, and accessible to all. Every autistic child is unique, every journey singular, but all deserve a chance to grow, to flourish, and to contribute to the world in their own way. Behind every diagnosis lies a precious life, an untapped potential, and a wealth that we, as a society, bear the responsibility to nurture.

Far from closing with absolute certainties, this book is an invitation: an invitation to keep searching, to explore new paths, and to unite our strengths in facing the challenges of autism. I deeply believe that progress will come from collaboration–between doctors, researchers, educators, and families, but also across disciplines and nations. Autism is not a solitary struggle. It is a shared responsibility, a call to build a world where no one is left behind.

If these pages have managed to shed light on certain aspects of autism, to offer new perspectives, or to provoke reflection, then they will have fulfilled their purpose. Yet they represent only a fragment of a larger, collective effort that will continue to grow. I dream of a future where autistic children will no longer be defined by their difficulties but by their talents and singularities. A future where their families will no longer feel isolated, judged, or forgotten, but supported at every stage of their journey.

I close these pages with my eyes set on tomorrow, convinced that the union of science, compassion, and determination can transform what seems insurmountable today into a horizon filled with infinite possibilities. Autism is not a fate carved in stone—it is a call to action, to hope, and above all, to love. Together, we can, and we must, build that future.

This book has no greater ambition than to spark deep reflection on autism, by exploring the multiple facets of a systemic and neuro-inflammatory condition. May we, together, open new perspectives to better understand, support, and transform this reality.

Further Reading

For further exploration, the following is a non-exhaustive selection of scientific articles corresponding to the various chapters of this book.

A Runaway Epidemiology: Autism in Numbers

Burd L, Fisher W, Kerbeshian J. A prevalence study of pervasive developmental disorders in North Dakota. J Am Acad Child Adolesc Psychiatry. 1987 Sep;26(5) :700-3.

Chiarotti F, Venerosi A. Epidemiology of Autism Spectrum Disorders : A Review of Worldwide Prevalence Estimates Since 2014. Brain Sci. 2020 May 1;10(5) :274.

Nevison CD. A comparison of temporal trends in United States autism prevalence to trends in suspected environmental factors. Environ Health. 2014 Sep 5;13 :73.

Beyond Diagnostic Labels: Rethinking Autism

Shreeya Gyawali, Bichitra Nanda Patra Trends in concept and nosology of autism spectrum disorder : A review in Asian Journal of Psychiatry Volume 40, February 2019, Pages 92-99

Poirier, N. & Forget, J. (1998). Les critères diagnostiques de l'autisme et du syndrome d'Asperger : similitudes et différences. Santé mentale au Québec, 23 (1), 130–148.

Janet K. Kern, David A. Geier, Lisa K. Sykes, Mark R. Geier, and Richard C. Deth Are ASD and ADHD a Continuum? A Comparison of Pathophysiological Similarities Between the Disorders in J Atten Disord. 2015 Sep ;19(9) :805-27

Reading Leo Kanner's First Article on Autism

Kanner L. Autistic disturbances of affective contact. Acta Paedopsychiatr. 1968;35(4) :100-36.

Autism's Inflammatory Nature

Irene Tsilioni, Arti B. Patela, Harry Pantazopoulos, Sabina Berretta, Pio Conti, Susan E. Leemane,, and Theoharis C. Theoharides IL-37 is increased in brains of children with autism spectrum disorder and inhibits human microglia stimulated by neurotensin in PNAS October 22, 2019 116 (43) 21659-21665

DA Rossignol and RE Frye A review of research trends in physiological abnormalities in autism spectrum disorders : immune dysregulation, inflammation, oxidative stress, mitochondrial dysfunction and environmental toxicant exposures in Molecular Psychiatry (2012) 17, 389–401

Benjamin Gesundheit , Joshua P Rosenzweig, David Naor, Bernard Lerer, Ditza A Zachor, Vaclav Procházka, Michal Melamed, Donald A Kristt, Abraham Steinberg, Cory Shulman, Paul Hwang, Gideon Koren, Asnat Walfisch, Jacob R Passweg, John A Snowden, Ryad Tamouza, Marion Leboyer, Dominique Farge-Bancel, Paul Ashwood Immunological and autoimmune considerations of Autism Spectrum Disorders in Journal of Autoimmunity Volume 44, August 2013, Pages 1-7

Mai Moaaza, Sara Youssrya ,Amr Elfatatry ,Mohammed Abd El Rahman Th17/Treg cells imbalance and their related cytokines (IL-17, IL-10 and TGF-β) in children with autism spectrum disorder in J Neuroimmunol . 2019 Dec 15;337 :577071

Dario Siniscalco, Stephen Schultz, Anna Lisa Brigida and Nicola Antonucci Inflammation and Neuro-Immune Dysregulations in Autism Spectrum Disorders in Pharmaceuticals 2018, 11, 5

AS Brown, A Sourander, S Hinkka-Yli-Saloma¨ki, IW McKeague, J Sundvall and H-M Surcel Elevated maternal C-reactive protein and autism in a national birth cohort in Molecular Psychiatry (2014) 19, 259–264

Turbé H, et al. Overview of prospects for inflammation pathways in autism spectrum disorders. Encéphale (2019)

Ashwood P, Krakowiak P, Hertz-Picciotto I, Hansen R, Pessah I, Van de Water J. Elevated plasma cytokines in autism spectrum disorders provide evidence of immune dysfunction and are associated with impaired behavioral outcome. Brain Behav Immun. 2011 Jan;25(1) :40-5

Melamed IR, Heffron M, Testori A, Lipe K. A pilot study of high-dose intravenous immunoglobulin 5% for autism : Impact on autism spectrum and markers of neuroinflammation. Autism Res. 2018 Mar;11(3) :421-433.

Mohammad Reza Khakzad, Maryam Javanbakht, Mohammad Reza Shayegan, Sina Kianoush, Fatemeh Omid, Maryam Hojati, Mojtaba Meshkat, The complementary role of high sensitivity C-reactive protein in the diagnosis and severity assessment of autism, Research in Autism Spectrum Disorders, Volume 6, Issue 3, 2012, Pages 1032-1037,

Ji Y, Azuine RE, Zhang Y, Hou W, Hong X, Wang G, Riley A, Pearson C, Zuckerman B, Wang X. Association of Cord Plasma Biomarkers of In Utero Acetaminophen Exposure With Risk of Attention-Deficit/Hyperactivity Disorder and Autism Spectrum Disorder in Childhood. JAMA Psychiatry. 2020 Feb 1;77(2) :180-189.

Nadeem R, Hussain T, Sajid H. C reactive protein elevation among children or among mothers' of children with autism during pregnancy, a review and meta-analysis. BMC Psychiatry. 2020 May 24;20(1) :251.

Autism and Autoimmunity: A Hidden Connection

Ashwood P, Wills S, Van de Water J. The immune response in autism : a new frontier for autism research. J Leukoc Biol. 2006 Jul;80(1) :1-15.

Ashwood P, Van de Water J. Is autism an autoimmune disease? Autoimmun Rev. 2004 Nov;3(7-8) :557-62.

Connery K, Tippett M, Delhey LM, Rose S, Slattery JC, Kahler SG, Hahn J, Kruger U, Cunningham MW, Shimasaki C, Frye RE. Intravenous immunoglobulin for the treatment of autoimmune encephalopathy in children with autism. Transl Psychiatry. 2018 Aug 10;8(1) :148.

Tianle Zou, Jun Liu, Xueying Zhang, Huilin Tang, Yiqing Song, Xuejun Kong, Autoantibody and autism spectrum disorder : A systematic review, Research in Autism Spectrum Disorders, Volume 75,2020,101568,

Autistic Regression: A Controversial Concept

Gomez-Fernandez A,de la Torre-Aguilar MJ, Gil-Campos M, Flores-Rojas K, Cruz-Rico MD, Martin-Borreguero P and Perez-Navero JL Children With Autism Spectrum Disorder With Regression Exhibit a Different Profile in Plasma Cytokines and Adhesion Molecules Compared to Children Without Such Regression.in Front. Pediatr. 6 :264 (2018)

Lucy Thompson, Christopher Gillberg, Sara Landberg, Anne-Katrin Kantzer,Carmela Miniscalco, Martina Barnevik Olsson Mats A. Eriksson and Elisabeth Fernell Autism With and Without Regression : A Two-Year Prospective Longitudinal Study in Two Population-Derived Swedish Cohorts in Journal of Autism and Developmental Disorders (2019) 49 :2281–2290

Brian D. Barger,Jonathan M. Campbell,Jaimi D. McDonough Prevalence and Onset of Regression within Autism Spectrum Disorders : A Meta-analytic Review in J Autism Dev Disord (2013) 43 :817–828

Williams, K, Brignell, A, Prior, M, Bartak, L, Roberts, J Regression in autism spectrum disorders,In Journal of Paediatrics and Child Health, Volume 51, Issue 1, pages 61–64, 2015

Katrina Williams,Amanda Brignell,Margot Prior,Lawrence Bartak and Jacqueline Roberts Regression in autism spectrum disorders in Journal of Paediatrics and Child Health (2015) 61–64

Gevezova M, Ivanov Z, Pacheva I, Timova E, Kazakova M, Kovacheva E, Ivanov I, Sarafian V. Bioenergetic and Inflammatory Alterations in Regressed and Non-Regressed Patients with Autism Spectrum Disorder. Int J Mol Sci. 2024 Jul 27;25(15) :8211.

The Genetic Puzzle of Autism

Zhou, J., Park, C.Y., Theesfeld, C.L. et al. Whole-genome deep-learning analysis identifies contribution of noncoding mutations to autism risk. Nat Genet 51, 973–980 (2019).

Sandin S, Lichtenstein P, Kuja-Halkola R, Hultman C, Larsson H, Reichenberg A. The Heritability of Autism Spectrum Disorder. JAMA. 2017 Sep 26;318(12) :1182-1184.

Bai D, Yip BHK, Windham GC, Sourander A, Francis R, Yoffe R, Glasson E, Mahjani B, Suominen A, Leonard H, Gissler M, Buxbaum JD, Wong K, Schendel D, Kodesh A, Breshnahan M, Levine SZ, Parner ET, Hansen SN, Hultman C, Reichenberg A, Sandin S. Association of Genetic and Environmental Factors With Autism in a 5-Country Cohort. JAMA Psychiatry. 2019 Oct 1;76(10) :1035-1043.

Genetic Polymorphisms and Autism: The Case of MTHFR

Marvin Boris,Allan Goldblatt, Joseph Galanko, S. Jill James.Association of MTHFR Gene Variants with Autism in Journal of American Physicians and Surgeons Volume 9 Number 4 Winter 2004

Rai V and Kumar P. Methylenetetrahydrofolate Reductase A1298C Polymorphism and Autism Susceptibility.in Austin J Autism & Relat Disabil. 2018; 4(1) : 1048.

Elif Funda Sener,Didem Behice Oztop Yusuf Ozkul MTHFR Gene C677T Polymorphism in Autism Spectrum Disorders in Genetics Research International Volume 2014,

Danhua Pu, Yiping Shen, and Jie Wu Association between MTHFR Gene Polymorphisms and the Risk of Autism Spectrum Disorders : A Meta-Analysis - Autism Res 2013

Autism and the Forgotten Organ: The Gut Microbiome

Sgritta M, Dooling SW, Buffington SA, Momin EN, Francis MB, Britton RA, Costa-Mattioli M. Mechanisms Underlying Microbial-Mediated Changes in Social Behavior in Mouse Models of Autism Spectrum Disorder. Neuron. 2019 Jan 16;101(2) :246-259.e6

Pulikkan J, Mazumder A, Grace T. Role of the Gut Microbiome in Autism Spectrum Disorders. Adv Exp Med Biol. 2019;1118 :253-269

Xu M, Xu X, Li J, Li F. Association Between Gut Microbiota and Autism Spectrum Disorder : A Systematic Review and Meta-Analysis. Front Psychiatry. 2019 Jul 17;10 :473.

Bezawada N, Phang TH, Hold GL, Hansen R. Autism Spectrum Disorder and the Gut Microbiota in Children : A Systematic Review. Ann Nutr Metab. 2020;76(1) :16-29.

Nalan H. Noğay, Chapter 12 - Effects of the Gut Microbiota on Autism Spectrum Disorder, In Handbook of Food Bioengineering, Diet, Microbiome and Health, Academic Press, 2018,Pages 347-368,

Bolte ER. Autism and Clostridium tetani. Med Hypotheses. 1998 Aug;51(2) :133-44.

Sharon G, Cruz NJ, Kang DW, Gandal MJ, Wang B, Kim YM, Zink EM, Casey CP, Taylor BC, Lane CJ, Bramer LM, Isern NG, Hoyt DW, Noecker C, Sweredoski MJ, Moradian A, Borenstein E, Jansson JK, Knight R, Metz TO, Lois C, Geschwind DH, Krajmalnik-Brown R, Mazmanian SK. Human Gut Microbiota from Autism Spectrum Disorder Promote Behavioral Symptoms in Mice. Cell. 2019 May 30;177(6) :1600-1618.e17

MacFabe DF. Enteric short-chain fatty acids : microbial messengers of metabolism, mitochondria, and mind : implications in autism spectrum disorders. Microb Ecol Health Dis. 2015 May 29;26 :28177

Sandler RH, Finegold SM, Bolte ER, Buchanan CP, Maxwell AP, Väisänen ML, Nelson MN, Wexler HM. Short-term benefit from oral vancomycin treatment of regressive-onset autism. J Child Neurol. 2000 Jul;15(7) :429-35.

The Brain-Gut Connection and the Vagus Nerve in Autism

Crystal T. Engineer, Seth A. Hays and Michael P. Kilgard Vagus nerve stimulation as a potential adjuvant to behavioral therapy for autism and other neurodevelopmental disorders in Journal of Neurodevelopmental Disorders (2017) 9 :20

Sheinkopf SJ, Levine TP, McCormick CEB, Puggioni G, Conradt E, Lagasse LL, Lester BM, Developmental Trajectories of Autonomic Functioning in Autism from Birth to Early Childhood, Biological Psychology (2019)

Stephen J. Sheinkopf, Todd P. Levine, Carolyn E.B. McCormick, Gavino Puggioni, Elisabeth Conradt, Linda L. Lagasse, Barry M. Lester Developmental Trajectories of Autonomic Functioning in Autism from Birth to Early Childhood in Neuron 101, 246–259 (2019)

Breit S, Kupferberg A, Rogler G and Hasler G : Vagus Nerve as Modulator of the Brain–Gut Axis in Psychiatric and Inflammatory Disorders. Front. Psychiatry 9 :44 (2018)

Perry EK, Lee ML, Martin-Ruiz CM, Court JA, Volsen SG, Merrit J, Folly E, Iversen PE, Bauman ML, Perry RH, Wenk GL.Cholinergic activity in autism : abnormalities in the cerebral cortex and basal forebrain.in Am J Psychiatry. 2001 Jul;158(7) :1058-66.

Autism as a Mitochondrial Disease

B. Fromenty Toxicité mitochondriale et métabolique des médicaments : mécanismes et conséquences au niveau du foie in Réanimation Volume 19, Issue 6, October 2010, Pages 552-567

Keren K. Griffiths and Richard J. Levy Evidence of Mitochondrial Dysfunction in Autism :

Biochemical Links, Genetic-Based Associations, and Non-Energy Related Mechanisms in Oxidative Medicine and Cellular Longevity Volume 2017

Shannon Rose, Dmitriy M. Niyazov, Daniel A. Rossignol,Michael Goldenthal,Stephen G. Kahler, Richard E. Frye Clinical and Molecular Characteristics of Mitochondrial Dysfunction in Autism Spectrum Disorder in Molecular Diagnosis & Therapy (2018) 22 :571–593

DA Rossignol and RE Frye Mitochondrial dysfunction in autism spectrum disorders : a systematic review and meta-analysis Mitochondrial dysfunction in autism spectrum disorders : a systematic review and meta-analysis in Molecular Psychiatry (2012) 17, 290–314

Katarina Babinská, Gabriela Repiská, Anna Gvozdjáková, Jarmila Kucharská, Lenka Vokálová , Mária Vidošovičová and Daniela Ostatníková What is the evidence of mitochondrial dysfunction in autism spectrum disorders? In Activitas Nervosa Superior Rediviva Volume 59 No. 2 2017

Richard E. Frye and Daniel A. Rossignol Mitochondrial dysfunction can connect the diverse medical symptoms associated with autism spectrum disorders in Pediatr Res. 2011 May ; 69

Mary Coleman and John P. Blass Autism and Lactic Acidosis in Journal of Autism and DevelopmentalDisorders, Vol. 15, No. l, 1985

Cecilia Giulivi,Yi-Fan Zhang,Alicja Omanska-Klusek, Catherine Ross-Inta, Sarah Wong, Irva Hertz-Picciotto, Mitochondrial Dysfunction in Autism in JAMA. 2010 December 1; 304(21) : 2389–2396

Abha Chauhan, Feng Gu, Musthafa M. Essa, Jerzy Wegiel, Kulbir Kaur, William Ted Brown, and Ved Chauhan, Brain region-specific deficit in mitochondrial electron transport chain complexes in children with autism in J Neurochem. 2011 April ; 117(2) : 209–220.

Guomei Tanga, Puri Gutierrez Rios, Sheng-Han Kuo, Hasan Orhan Akman, Gorazd Rosoklija, Kurenai Tanji , Andrew Dwork , Eric A. Schon, Salvatore DiMauro, James Goldman , and David Sulzer, Mitochondrial abnormalities in temporal lobe of autistic brain in Neurobiol Dis. 2013 June ; 54 : 349–361

Suzanne Goh,, Zhengchao Dong,, Yudong Zhang,, Salvatore DiMauro, and Bradley S.Peterson, Brain imaging evidence that mitochondrial dysfunction is a neurobiological subtype of Autism Spectrum Disorder in JAMA Psychiatry. 2014 June ; 71(6) : 665–671

Rossignol DA and Frye RE (2014) Evidence linking oxidative stress, mitochondrial dysfunction, and inflammation in the brain of individuals with autism. Front. Physiol. 5 :150

Schultz ST, Klonoff-Cohen HS, Wingard DL, Akshoomoff NA, Macera CA, Ji M. Acetaminophen (paracetamol) use, measles-mumps-rubella vaccination, and autistic disorder : the results of a parent survey. Autism. 2008 May;12(3) :293-307.

Becker KG, Schultz ST. Similarities in features of autism and asthma and a possible link to acetaminophen use. Med Hypotheses. 2010 Jan;74(1) :7-11.

Bührer C, Endesfelder S, Scheuer T, Schmitz T. Paracetamol (Acetaminophen) and the Developing Brain. Int J Mol Sci. 2021 Oct 15;22(20) :11156.

Schultz ST, Gould GG. Acetaminophen Use for Fever in Children Associated with Autism Spectrum Disorder. Autism Open Access. 2016 Apr;6(2) :170.

Good P. Did acetaminophen provoke the autism epidemic? Altern Med Rev. 2009 Dec;14(4) :364-72.

Autism and Mast Cell Activation

Theoharides TC, Tsilioni I, Patel AB, Doyle R. Atopic diseases and inflammation of the brain in the pathogenesis of autism spectrum disorders. Transl Psychiatry. 2016 Jun 28;6(6) :e844.

Theoharides TC. Is a subtype of autism an allergy of the brain? Clin Ther. 2013 May;35(5) :584-91.

Theoharides TC. Mast cells : the immune gate to the brain. Life Sci. 1990;46(9) :607-17.

Theoharides TC, Kavalioti M, Martinotti R. Factors adversely influencing neurodevelopment. J Biol Regul Homeost Agents. 2019 Nov-Dec;33(6) :1663-1667.

Autism and the Thyroid Connection

Darios Getahun, Steven J. Jacobsen, Michael J. Fassett, Deborah A. Wing , Anny H. Xiang, Vicki Y. Chiu and Morgan R. Peltier Association between maternal hypothyroidism and autism spectrum disorders in children Pediatric RESEARCH Volume 83 Number 3 March 2018

R. E. Frye R. Wynne S. Rose J. Slattery L. Delhey M. Tippett S. G. Kahler S. C. Bennuri S. Melnyk J. M. Sequeira E. V. Quadros Thyroid dysfunction in children with autism spectrum disorder is associated with folate receptor α autoimmune disorder in J. Neuroendocrinol. 29(3),

Getahun D, Jacobsen SJ, Fassett MJ, Wing DA, Xiang AH, Chiu VY, Peltier MR. Association between maternal hypothyroidism and autism spectrum disorders in children. Pediatr Res. 2018 Mar;83(3) :580-588

Carnitine and Autism: A Hidden Connection

Caroline Demarquoy, Jean Demarquoy Autism and carnitine : A possible link in World J Biol Chem 2019 January 7; 10(1) : 7-16

Kępka A, Ochocińska A, Chojnowska S, Borzym-Kluczyk M, Skorupa E, Knaś M, Waszkiewicz N. Potential Role of L-Carnitine in Autism Spectrum Disorder. J Clin Med. 2021 Mar 13;10(6) :1202.

Autism and Calcium: A Hidden Connection

Rachel L. Nguyen, Yuliya V. Medvedeva, Tejasvi E. Ayyagari, Galina Schmunk, John Jay Gargus , Intracellular calcium dysregulation in autism spectrum disorder : An analysis of converging organelle signaling pathways. Bbamcr (2018),

Schmunk, G. et al. High-throughput screen detects calcium signaling dysfunction intypical sporadic autism spectrum disorder. Sci. Rep. 7, 40740; (2017).

Vitamin D: The Missing Piece of the Autism Puzzle

L. MÁČOVÁ, M. BIČÍKOVÁ , D. OSTATNÍKOVÁ , M. HILL, L. STÁRKA, Vitamin D, Neurosteroids and Autism in Physiol. Res. 66 (Suppl. 3) : S333-S340, 2017

John Jacob Cannell Vitamin D and autism, what's new? In Rev Endocr Metab Disord (2017) 18 :183–193

Rhonda P. Patrick and Bruce N. Ames Vitamin D hormone regulates serotonin synthesis. Part 1 : relevance for autism in the FASEB Journal Vol. 28 June 2014

Alerie Guzman de la Fuente, Oihana Errea, Peter van Wijngaarden, Ginez A. Gonzalez, Christophe Kerninon, Andrew A. Jarjour, Hilary J. Lewis, Clare A. Jones, Brahim Nait-Oumesmar, Chao Zhao, Jeffrey K. Huang, Charles ffrench-Constant, and Robin J.M. Franklin Vitamin D receptor–retinoid X receptor heterodimer signaling regulates oligodendrocyte progenitor cell differentiation in J. Cell Biol. Vol. 211 No. 5 975–985.

Petruzzelli MG, Marzulli L, Margari F, De Giacomo A, Gabellone A, Giannico OV, Margari L. Vitamin D Deficiency in Autism Spectrum Disorder : A Cross-Sectional Study. Dis Markers. 2020 Sep 18;2020 :9292560.

Song L, Luo X, Jiang Q, Chen Z, Zhou L, Wang D, Chen A. Vitamin D Supplementation is Beneficial for Children with Autism Spectrum Disorder : A Meta-analysis. Clin Psychopharmacol Neurosci 2020;18 :203-213.

The Role of GABA and Glutamate Systems in Autism

Lemonnier E, Villeneuve N, Sonie S, Serret S, Rosier A, Roue M, Brosset P, Viellard M, Bernoux D, Rondeau S, Thummler S, Ravel D, Ben-Ari Y. Effects of bumetanide on neurobehavioral function in children and adolescents with autism spectrum disorders. Transl Psychiatry. 2017 Mar 14;7(3) :e1056.

Bhandari R, Varma M, Rana P, Dhingra N, Kuhad A. Taurine as a potential therapeutic agent interacting with multiple signaling pathways implicated in autism spectrum disorder (ASD) : An in-silico analysis. IBRO Neurosci Rep. 2023 Sep 2;15 :170-177.

Marotta R, Risoleo MC, Messina G, Parisi L, Carotenuto M, Vetri L, Roccella M. The Neurochemistry of Autism. Brain Sci. 2020 Mar 13;10(3) :163.

Autism and Microglia

Tomoyuki Takano Role of Microglia in Autism : Recent Advances Dev Neurosci2015;37 :195–202

Jung Won Kim , Ji Yeon Hong , Seung Min Bae Microglia and Autism Spectrum Disorder : Overview of Current Evidence and Novel Immunomodulatory Treatment Options, in Clinical Psychopharmacology and Neuroscience 2018;16(3) :246-252

Rodney W. Johnson From Inflammation to Sickness and Cognitive Dysfunction : When the Immune System Subjugates the Brain in The 114th Abbott Nutrition Research Conference "Cognition and nutrition" April 8-9, 2013 Columbus, Ohio, USA

Xu, ZX., Kim, G.H., Tan, JW. et al. Elevated protein synthesis in microglia causes autism-like synaptic and behavioral aberrations. Nat Commun 11, 1797 (2020).

The Future Perspective: The Cell Danger Response Theory

Naviaux RK. Perspective : Cell danger response Biology-The new science that connects environmental health with mitochondria and the rising tide of chronic illness. Mitochondrion. 2020 Mar;51 :40-45.

Naviaux RK, Zolkipli Z, Wang L, Nakayama T, Naviaux JC, et al. (2013) Antipurinergic Therapy Corrects the Autism-Like Features in the Poly(IC) Mouse Model. PLoS ONE 8(3) : e57380.

JC Naviaux, MA Schuchbauer, K Li, L Wang, VB Risbrough, SB Powell and RK Naviaux, Reversal of autism-like behaviors and metabolism in adult mice with single-dose antipurinergic therapy in Translational Psychiatry (2014) 4, e400;

Autism and the Kynurenine Pathway

Santana-Coelho D. Does the kynurenine pathway play a pathogenic role in autism spectrum disorder? Brain Behav Immun Health. 2024 Aug 6;40 :100839.

Mega-Analysis and Clinical Research on Autism

Emily Simonoff Commentary : Randomized controlled trials in autism spectrum disorder : state of the field and challenges for the future in Journal of Child Psychology and Psychiatry 59 :4 (2018), pp 457–459

Translational Medicine: A Therapeutic Hope for Autism

HUANG, Xin-en. (2016). Relationship among Translational Medicine, Evidence-Based Medicine and Precision Medicine. Journal of International Translational Medicine.

Fecal Transplants and Autism: What the New Science—and New Rules—Really Say

Dae-Wook Kang, James B.Adams, Devon M. Coleman, Elena L. Pollard, Juan Maldonado, Sharon McDonough-Means, J.GregoryCaporaso, Rosa Krajmalnik-Brown Long-term benefit of Microbiota Transfer Therapy on autism symptoms and gut microbiota in Scientific Reports | (2019) 9 :5821

Kang et al, Microbiota Transfer Therapy alters gut ecosystem and improves gastrointestinal and autism symptoms : an open-label study in Kang et al. Microbiome (2017) 5 :10

Agence nationale de sécurité du médicament et des produits de santé (France) : La transplantation de microbiote fécal et son encadrement dans les essais cliniques - Novembre 2016 (Actualisation de la version de juin 2015)

Stem Cell Therapy: Between Hope and Illusion

Siniscalco D, Kannan S, Semprún-Hernández N, Eshraghi AA, Brigida AL, Antonucci N. Stem cell therapy in autism : recent insights. Stem Cells Cloning. 2018 Oct 23;11 :55-67

Future Perspectives: Transcranial Magnetic Stimulation

Barahona-Corrêa JB, Velosa A,Chainho A, Lopes R and Oliveira-Maia AJ (2018) Repetitive Transcranial Magnetic Stimulation for Treatment of Autism Spectrum Disorder : A Systematic Review and Meta-Analysis in Front. Integr. Neurosci. 12 :27

Antoni Valero-Cabré,Alvaro Pascual-Leone, and Olivier A. Coubard La stimulation magnétique transcrânienne (SMT) dans la recherche fondamentale et clinique en neuroscience in Dev Neurosci 2015;37 :195–202

Advocacy for Early Autism Diagnosis

Elder JH, Kreider CM, Brasher SN, Ansell M.Clinical impact of early diagnosis of autism on the prognosis and parent–child relationships in Psychology Research and Behavior Management 2017 :10 283–292

For a Truly Inclusive School

Mazzer, Marzia. "Inclusion of Students with Autism Spectrum Disorders in Italian Schools : From Theory to Practice." Open Journal of Social Sciences 03 (2015) : 91-96.

The author, Dr. Othmane Boumaalif, warmly welcomes readers to continue the dialogue and exchange of ideas sparked by this book. He believes that understanding grows through collective reflection and shared experience. For questions, comments, or to contribute to this ongoing conversation, readers are invited to reach out directly at **boumaalif@gmail.com**

Dépôt légal octobre 2025

Librinova

www.librinova.com

www.ingramcontent.com/pod-product-compliance
Lightning Source LLC
LaVergne TN
LVHW100310210926
848480LV00003B/1086
* 9 7 9 1 0 4 0 5 9 3 0 4 1 *